INTRODUCTION TO
REAL ANALYSIS

INTRODUCTION TO REAL ANALYSIS

An Educational Approach

William C. Bauldry

Appalachian State University

A JOHN WILEY & SONS, INC., PUBLICATION

Library of Congress Cataloging-in-Publication Data:

Bauldry, William C.
  Introduction to real analysis : an educational approach / William C. Bauldry.
    p. cm.
  Includes bibliographical references and index.
  ISBN 978-0-470-37136-7 (cloth)
  1. Mathematical analysis—Textbooks. 2. Functions—Textbooks. I. Title.
  QA300.B38 2009
  515—dc22   2009013335

Printed in the United States of America.

10 9 8 7 6 5 4 3 2 1

To my sons,
Shawn and Mike,
two of the finest.

CONTENTS

PREFACE

In most sciences one generation tears down what another has built, and what one has established another undoes. In Mathematics alone each generation adds a new story to the old structure.

— Hermann Hankel

Elementary calculus students develop intuitive concepts of limits, derivatives, and integrals. While adequate for most, those who teach calculus need a much deeper understanding. The purpose of this book is to present an introductory course in analysis, rebuilding intuitive understandings with more rigorous foundations, with emphases chosen for the practicing secondary teacher. The text is based on a one-semester, four-credit-hour, graduate course in our Masters in Mathematics with a concentration in secondary education. Since we wish to cover a significant amount of material in one semester, some topics will necessarily be given short attention and others are not covered. This reduction allows us to introduce Lebesgue measure and integration. The last chapter presents a set of special topics with ties to elementary calculus that can be used as student projects. These topics can be studied in more depth if time permits.

Epigraph. From *Die Entwicklung der Mathematik in den letzten Jahrhunderten,* 2nd ed., Antrittsvorlesung, (Tübingen, 1889), p. 25. Quoted in Moritz's *On Mathematics and Mathematicians* (Moritz, 1958, p. 14).

To The Student

Why study real analysis? There are many reasons. Real analysis is a fascinating and elegant area containing many deep results that are important throughout mathematics. Calculus, which grew to become real analysis, is considered one of the crowning intellectual achievements of humankind with roots as deep as Archimedes. For a very practical answer, most elementary calculus instruction now occurs in secondary schools. For high school faculty to teach calculus well requires, in the words of Liping Ma (1999), a "profound understanding of fundamental mathematics." The underlying fundamental mathematics of calculus forms the core of real analysis. Edwina Michener begins her article "Understanding Understanding Mathematics" (Michener, 1978) with

> When a mathematician says he understands a mathematical theory, he possesses much more knowledge than that which concerns the deductive aspects of theorems and proofs. He knows about examples and heuristics and how they are related. He has a sense of what to use and when to use it, and what is worth remembering. He has an intuitive feeling for the subject, how it hangs together, and how it relates to other theories. He knows how not to be swamped by details, but also to reference them when he needs them.

Our goal is to help build that foundation.

We begin with a quick overview of first-year calculus in Chapter 1, pausing to examine problems illustrating links and potential difficulties. Then we study real analysis in Chapter 2 starting with basic properties of the real numbers and going through limits to integration to series of functions in a natural progression. As we carefully prove theorems along the way, we need to focus on the results and proofs that help to lead us to a deeper understanding of basics. That's not to say that we'll only look at elementary analysis—Chapter 3 is an introduction to Lebesgue theory, analysis from a very advanced viewpoint. Understanding the basics of Lebesgue's approach to integration and measure provides a superior foundation for elementary real analysis and offers a view toward more advanced topics. The last chapter is a selection of special topics that lead to other areas of study. These topics form the germs of excellent projects. The appendices serve as references. Appendix A lists the definitions and theorems of elementary analysis, Appendix B offers a very brief timeline placing Newton and Leibniz in perspective, and Appendix C is a collection of projects. Appendix C also has pointers to other sources of projects, both for real analysis and for calculus. Pay attention to the historical development; history helps us see the connections among the themes of analysis.

As you work through the text, remember to concentrate not only on the idea at hand, but also on how it fits into the larger picture of analysis. Keep in mind Polya's advice from *How to Solve It*, "A well-stocked and well-organized body of knowledge is an asset to the problem solver." As an aid to organizing your personal calculus and analysis framework, build a concept map for each major topic. A sample concept map picturing "derivative" is available at

www.mathsci.appstate.edu/~wmcb/IA/DerivativeMap/Derivative.html

To The Instructor

Teaching a real analysis course in a graduate program for teachers presents special challenges to the instructor. Students' backgrounds are extremely varied: Many undergraduate programs no longer require advanced calculus, many students are in-service teachers who may be several years past their undergraduate studies, other students may be on alternative pathways to teaching licensure and have minimal undergraduate mathematics training. Often the class is offered in an abbreviated summer semester. These factors combine to become significant issues for the professor of introductory analysis. We attempt to address these aspects in several ways. Beginning with an overview of typical calculus material helps to refresh students' memory and helps them to refocus in a conceptual, theoretical framework, preparing students to study analysis. Then following a path through introductory analysis that recapitulates the earlier calculus presentation helps students to see development and depth relative to their needs as mathematics instructors.

The history of calculus and analysis is woven throughout this text. Historical context helps a student to place the mathematics and to better understand the imperative of developments. This philosophy is best expressed by Torkil Heide (1996):

> The history of mathematics is not just a box of paints with which one can make the picture of mathematics more colourful, to catch interest of students at their different levels of education; it is a part of the picture itself. If it is such an important part that it will give a better understanding of what mathematics is all about, if it will widen horizons of learners, maybe not only their mathematical horizons . . . then it must be included in teaching.

Appalachian's four-credit "Analysis for Teachers" graduate course can be offered in either a standard fourteen-week semester plus final exams or an accelerated four-week summer session. Our prerequisites are admission to the mathematics graduate program and the equivalent of an undergraduate mathematics major, including at least three credits of advanced calculus.

A typical syllabus for a fourteen-week semester with four class days per week appears in Table P.1, Sample Syllabus. The material in Chapter 1 is covered quickly, the time is spent mainly with examples and problems illustrating the topics. Some problems are chosen to lead to pitfalls. Chapter 2 introduces real analysis. Theorems and their proofs are carefully considered. The proofs shown are chosen for techniques and accessibility. Chapter 2 is where the instructor most needs to adjust timing based on individual classes. Chapter 3 is an introduction to Lebesgue theory from the perspective of building towards convergence results in relation to integration. Chapter 4 presents several special topics that relate to or extend standard elementary calculus concepts. For example, the first section of Chapter 4 weaves together numerical derivatives from data, data transformation, and the logistic model. The number of special topics chosen and the depth of coverage given allow an instructor a good deal of flexibility. Student presentations, as either individuals or groups, finish the course. Past presentations have ranged from historical accounts to pedagogical investigations to special topics such as Vitali's nonmeasurable set. Students uniformly enjoy and learn a great deal from these projects. Student presentations have always been an integral part of the course, even when under severe time-pressure.

Week	Day	Topic	Week	Day	Topic
1	1	Course Intro, §1.1	8	29	§2.6
	2	§1.2		30	§2.6
	3	§1.2		31	Problem Day
	4	§1.3		32	Exam 2
2	5	§1.3	9	33	§3.1
	6	§1.4		34	§3.1
	7	§1.4		35	§3.2
	8	§1.5, Quiz 1		36	§3.2
3	9	§1.5	10	37	§3.2
	10	§1.6		38	§3.3
	11	§1.6		39	§3.3
	12	Problem Day		40	§3.3, Quiz 4
4	13	Exam 1	11	41	§3.4
	14	§2.1		42	Problem Day
	15	§2.1		43	Exam 3
	16	§2.2		44	§4.1
5	17	§2.2	12	45	§4.1
	18	§2.2		46	§4.1
	19	§2.3		47	§4.2
	20	§2.3, Quiz 2		48	§4.2
6	21	§2.3	13	49	§4.2
	22	§2.4		50	§4.x
	23	§2.4		51	§4.x, Quiz 5
	24	§2.4		52	Student presentations
7	25	§2.4	14	53	Student presentations
	26	§2.5		54	Student presentations
	27	§2.5		55	Student presentations
	28	§2.6, Quiz 3		56	Student presentations
			15		Final Exam & Project Papers Due

Table P.1 Sample Syllabus

While total classroom time is the same in our four-week summer sessions, time for student reflection is extremely shortened, so we spend more minutes per day on each topic. The special topics of Chapter 4 may be replaced by student presentations. In cases of extreme time problems, Riemann-Stieltjes integration reluctantly disappeared.

WILLIAM C. BAULDRY
BauldryWC@appstate.edu

Boone, North Carolina
May, 2009

ACKNOWLEDGMENTS

I appreciate the thoughtful advice of a collection of anonymous reviewers; they have helped shaped the text and quash errors that crept into the manuscript. Thoughts and corrections are always welcome via email. Many thanks go to Wade Ellis and to Rich West for sharing their considerable wisdom over the years and for helping me keep students first in mind. I am grateful to the the the students of MAT 5930, Analysis for Teachers, for struggling with me while I developed preliminary versions of this text. Also, thanks go to the Department of Mathematical Sciences of Appalachian State University. The support of my colleagues has been invaluable; foremost among them are Jeff Hirst, Witold Kosmala, and Greg Rhoads, whose many conversations about analysis and pedagogy have strongly influenced my views and teaching. The staff at Wiley has always been cheerful and extremely helpful—they are the best. My thanks to Amy Hendrickson of TeXnology, a master of LaTeX—she makes it work. Special thanks go to my editor Susanne Steitz-Filler, a consummate professional, to Jackie Palmieri, Editorial Coordinator par excellence, and to Lisa Van Horn, who makes production extremely easy. Finally, I wish to thank my wife, Sue, for encouragement, and especially for putting up with "yet another writing project."

W. C. B.

CHAPTER 1

ELEMENTARY CALCULUS

Introduction

We begin our studies by reviewing the topics that form the standard core of a first-year college calculus or an AP calculus (AB or BC) class. Precalculus topics such as the real numbers and functions are assumed to be well known. Our plan in this chapter is to briefly describe the concept, placing the idea historically, and then immediately move to definitions and results, followed by examples and problems. The problems will be chosen to illustrate both examples and counterexamples and to lead to further thought.

1.1 PRELIMINARY CONCEPTS

We will take a number of concepts as previously understood. We are familiar with the properties of the standard number systems, the natural numbers $\mathbb{N}$, integers $\mathbb{Z}$, rational numbers $\mathbb{Q}$, and the real numbers $\mathbb{R}$. For two real numbers $a < b$, the *open interval* (a, b) is the set $\{x \in \mathbb{R} \mid a < x < b\}$. Similarly, the *closed interval* is

Introduction to Real Analysis. By William C. Bauldry
Copyright © 2009 John Wiley & Sons, Inc.

$[a, b] = \{x \in \mathbb{R} \mid a \leq x \leq b\}$. We do not use [or] for infinite intervals since ∞ is not a real number. An interval may be half-open by including only one endpoint.

A *function* f is a set of ordered pairs such that no two different pairs have the same first coordinate. For a function f, the set of first coordinates of the ordered pairs is called the *domain*; we write $\mathrm{dom}(f)$. The set of second coordinates of the function is called the *image*. The *range* of a function is a superset of the image. If X is the domain of the function f and Y is the range, we write $f : X \to Y$ and say "f maps X into Y." For $f : X \to Y$ with $A \subseteq X$ and $B \subseteq Y$, we set $f(A) = \{f(x) \in Y \mid x \in A\}$ and $f^{-1}(B) = \{x \in X \mid f(x) \in B\}$. Note that writing $f^{-1}(B)$ does not imply that f^{-1} is a function. A function is *injective* or *one-to-one* if and only if whenever $f(x_1) = f(x_2)$ then we must have $x_1 = x_2$. A function is *surjective* or *onto* if and only if for each $y \in Y$ there must be an $x \in X$ with $f(x) = y$. A *bijective* function is both one-to-one and onto. If f is bijective, then f^{-1} is a function. If f is not one-to-one, then f^{-1} cannot be a function.

The absolute value

$$|x| = \begin{cases} x & x \geq 0 \\ -x & x < 0 \end{cases}$$

gives the distance from x to 0; thus $|x - y|$ gives the distance from x to y. The following inequalities hold for any real numbers a, a_i, b, and b_i.

- $|a| \geq 0$

- $|a| \leq b$ if and only if $-b \leq a \leq b$

- $|a \cdot b| = |a| \cdot |b|$

- The *triangle inequality* $|a + b| \leq |a| + |b|$

- $\big||a| - |b|\big| \leq |a - b|$

- The *Cauchy-Bunyakovsky-Schwarz inequality*

$$\left[\sum_{i=1}^{n} a_i b_i\right]^2 \leq \left[\sum_{i=1}^{n} a_i^2\right] \cdot \left[\sum_{i=1}^{n} b_i^2\right]$$

- The *Minkowski inequality*

$$\sqrt{\sum_{i=1}^{n} (a_i + b_i)^2} \leq \sqrt{\sum_{i=1}^{n} a_i^2} + \sqrt{\sum_{i=1}^{n} b_i^2}$$

An *elementary function* is built from a finite combination of operations $(+, -, \times, \div, \sqrt[n]{\ }, \circ)$ applied to polynomials, exponentials, logarithms, or trigonometric or inverse trigonometric functions. If we allow complex numbers, then the trigonometric functions and their inverses are included as exponentials and logarithms.

For proofs and further information, refer to any advanced calculus text or outline; e.g., Wrede & Spiegel (2002) or Farand & Poxon (1984). We will revisit the properties of the real numbers more formally in Chapter 2.

1.2   LIMITS AND CONTINUITY

Even though studied first in calculus classes, historically, limits and continuity were developed formally as concepts well after integrals and derivatives were established (see the calculus timeline in Appendix B). The need to put rigorous foundations under the calculus led to the modern definitions of limit first given by Bolzano and Cauchy. Both Newton and Leibniz wrestled with justifying limits in their computations without solving the problem. Newton, following the lead of Fermat, used what we would call infinitesimals and let them be zero at the end of a computation, interpreting all constructions geometrically. Leibniz used equivalent differentials but worked from an arithmetic point of view. [See, e.g., Boyer (1959, Chapter V) or Burton (2007, Chapter 8).] The dispute over credit for inventing calculus between Newton, Leibniz, and their followers was very unfortunate. Bardi gives an account of the controversy in *The Calculus Wars* (Bardi, 2006) with a thorough historical perspective.

Limits of Functions

We'll begin by looking at intuitive forms of the definitions. Our first version will be taken from Cauchy's seminal 1821 text, *Cours d'Analyse* (Cauchy, 1821). These statements would be encountered in an elementary calculus course. Through problems, we move from an elementary definition to difficulties and challenges that will, in turn, lead us to Weierstrass's ϵ-δ definition of the limit.

Our first experience in calculus was to see limits numerically and graphically, often with the heuristics "are the function values coming together" or "do the two sides of the graph appear to connect?" These intuitive approaches work well for elementary calculus but are not sufficient for a deeper understanding and may even lead to misperceptions.

Definition 1.1 (Cauchy) *When the successive values attributed to a variable approach indefinitely a fixed value so as to end by differing from it by as little as one wishes, this last is called the* limit *of all the others.*

Or, in modern terms, we write

$$\lim_{x \to a} f(x) = L$$

if and only if we can make values of $f(x)$ arbitrarily close to L by taking x sufficiently close but not equal to a.

Before we try to compute limits, we need to know that there is at most one answer.

Theorem 1.1 *If $f(x)$ has a limit at $x = a$, that limit is unique.*

Now it's time to do some computations.

■ **EXAMPLE 1.1**

1. Determine the value of $\lim_{x \to 1} 5x - 2$.
   Either a table of values or a graph quickly leads us to the answer 3. *Do both!*

2. Let U be the *unit step function*

$$U(x) = \begin{cases} 1 & x > 0 \\ 0 & x \leq 0 \end{cases}$$

Find $\lim_{x \to 0} U(x)$.
Again using a table or graph, we see this limit does not exist. Verify this! ∎

The simple examples above help us to begin to gain intuition on limits. We can see whether function values "approach indefinitely a fixed value" with a table or with graphs. However, we quickly run into trouble with expressions that are a little more complex. Try generating tables and graphs for the limits in Example 1.2. No algebraic simplifications allowed!

■ **EXAMPLE 1.2**

1. What is the value of
$$\lim_{x \to 2} \frac{x - 2}{x^2 - 4}$$

2. What is the value of
$$\lim_{x \to 2} \frac{x + 2}{x^2 - 4}$$

3. Evaluate
$$\lim_{y \to 0} \frac{\sqrt{y^2 + 16} - 4}{y^2}$$

4. Calculate
$$\lim_{z \to 0} \sqrt[z]{1 + z}$$

∎

Several of the limits above present difficulties that can be easily overcome with algebra. The natural reduction in the first two limits in Example 1.2 reduces the expressions to simple forms. Rationalizing the numerator of the third limit again leaves an easier form. We need theorems for the algebra of limits. The last limit can be tamed by applying logarithms, but we'll need *continuity* to handle that calculation; we'll return to this limit later.

Theorem 1.2 (Algebra of Limits) *Suppose that $f(x)$ and $g(x)$ both have finite limits at $x = a$ and let $c \in \mathbb{R}$. Then*

- $\lim_{x \to a} c f(x) = c \lim_{x \to a} f(x)$

- $\lim_{x \to a} f(x) \pm g(x) = \lim_{x \to a} f(x) \pm \lim_{x \to a} g(x)$

- $\lim\limits_{x \to a} f(x) \cdot g(x) = \lim\limits_{x \to a} f(x) \cdot \lim\limits_{x \to a} g(x)$

- *if* $\lim\limits_{x \to a} g(x) \neq 0$, *then* $\lim\limits_{x \to a} \dfrac{f(x)}{g(x)} = \dfrac{\lim\limits_{x \to a} f(x)}{\lim\limits_{x \to a} g(x)}$

Our challenge now is to understand how to verify that the answers in the examples are correct. The algebra of limits lets us reduce to simpler forms but doesn't establish particular instances. We could formally show that $\lim_{x \to a} x = a$ and then apply the algebraic rules to handle rational functions at any point in their domain. Nevertheless, this technique would leave a lot of functions unresolved; i.e., all *transcendental functions*. A transcendental function is one that does not satisfy a polynomial equation whose coefficients are themselves polynomials. For example, sines and logarithms are transcendental.

A very useful theorem for computing limits is based on controlling an expression by finding upper and lower bounding functions and sandwiching the desired limit between.

Theorem 1.3 (Sandwich Theorem) *Suppose that* $g(x) \leq f(x) \leq h(x)$ *for all* x *in an open interval containing* a. *If both* $\lim_{x \to a} g(x) = \lim_{x \to a} h(x) = L$, *then* $\lim_{x \to a} f(x) = L$.

Let's use the sandwich theorem to find an interesting limit.

■ **EXAMPLE 1.3**

- Determine
$$\lim_{\phi \to 0} \frac{1 - \cos(\phi)}{\phi^2}$$

Using graphs, we are led to the inequality

$$\frac{1}{2} - \phi^2 \leq \frac{1 - \cos(\phi)}{\phi^2} \leq \frac{1}{2}$$

The sandwich theorem now easily gives the value of the limit as $1/2$. *Draw the graphs!* ■

Even though the sandwich theorem is very powerful, the last limit of Example 1.2 resists algebra and it's not easy to find bounding functions for $\sqrt[z]{1 + z}$. Our intuitive definition doesn't offer guidance for the task of computing limits. What we really need is to formalize what the phrases "arbitrarily close" and "sufficiently close" mean. The "ϵ-δ definition" we use today is attributed to Weierstrass; we state it in a form appropriate to a basic calculus course.

Definition 1.2 (Weierstrass) *Let* f *be defined on an open interval* $\mathcal{I}$ *containing* a, *except possibly at* a *itself. Then*

$$\lim_{x \to a} f(x) = L$$

if and only if for every $\epsilon > 0$ there is a $\delta > 0$ such that for every x in the interval $\mathcal{I}$, if $0 < |x - a| < \delta$, then $|f(x) - L| < \epsilon$.

Another advantage of this definition is that it easily splits into one sided limits. Often, it is easier to analyze a function from one side of a point than both sides simultaneously.

Definition 1.3 *Let f be defined on an open interval $\mathcal{I}$ containing a, except possibly at a itself. Then the* limit from the left *of f is M and the* limit from the right *is N,*

$$\lim_{x \to a-} f(x) = M \quad and \quad \lim_{x \to a+} f(x) = N$$

respectively, if and only if for every $\epsilon > 0$ there is a $\delta > 0$ such that for every x in the interval $\mathcal{I}$, if $0 < a - x < \delta$, then $|f(x) - M| < \epsilon$ from the left and if $0 < x - a < \delta$, then $|f(x) - N| < \epsilon$ from the right.

The limits from the left and right must match for the limit to exist.

Theorem 1.4 *The limit of f as x approaches a exists and equals L if and only if the limits from both the left and the right exist and are each equal to L.*

We can now give a demonstration that the unit step function has no limit at $a = 0$. Choose $\epsilon = 1/2$. Showing that no δ suffices is left to the exercises. Compare with showing that the limit doesn't exist since the left- and right-hand limits are different.

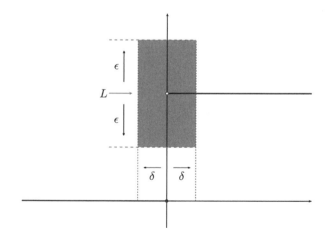

Figure 1.1   An ϵ-δ Box for the Unit Step Function

Verifying the last limit of Example 1.2 by the definition would be quite hard. We need to combine the concepts of limit and continuity to find that limit.

Continuity

The main impetus for studying limits in the beginning of a calculus course is to set the stage for continuity and for computations giving derivatives and integrals. Newton and Leibniz only worked with expressions that were essentially continuous functions. While Leibniz was the first to use the term *function*, mathematicians up through Euler's time considered a function as something that could be written as a single expression, thus eliminating even simple types such as step functions (Burton, 2007, p. 611). Today we have a broader conception of function but still use a narrower view in beginning classes. Current calculus texts follow the lead, either explicitly or implicitly, of Granville, Smith, and Longley (1911, p. 16): "In this book we shall only deal with functions which are in general continuous, that is, continuous for all values of x, with the possible exception of certain isolated values." Since most "calculus-style" functions are quite simple, the common heuristic for describing a continuous function is "graphing without lifting the pencil." Though intuitively appealing, this rubric quickly breaks down when considering more general functions and, as we'll see in Chapter 2, is actually misleading.

Let's start with a standard definition.

Definition 1.4 *The function f is* continuous *at $x = a$ if and only if*

$$\lim_{x \to a} f(x) = f(a)$$

Definition 1.4 combines three conditions into a single limit statement.

1. "$= f(a)$" requires that a belongs to the domain of f; i.e., f has a value at a.

2. "$\lim_{x \to a} f(x) =$" requires that the limit exists.

3. "$\lim_{x \to a} f(x) = f(a)$" requires that the value of the limit match the value of the function at a.

Creating examples violating each condition listed above is left to the exercises.

Since the ϵ-δ definition of limit specifies that f must be defined on an interval containing the point in question, we have eliminated isolated points of f's domain from our discussion of limits and hence from consideration as points of continuity.

■ **EXAMPLE 1.4**

1. Show that linear functions $f(x) = mx + b$ are continuous everywhere.
   Simple application of the definition.

2. Verify that the rational functions $r(x) = (ax^2 + bx + c)/(a'x^2 + b'x + c')$ are continuous wherever defined.
   Another simple application of the definition.

3. Let

$$g(\theta) = \begin{cases} \sin(\theta)/\theta x & \neq 0 \\ \alpha & x = 0 \end{cases}$$

Determine a value for α that makes g continuous at $\theta = 0$, if possible. *We need to find* $\lim_{\theta \to 0} g(\theta)$. *First, graph it!* ∎

A function that is continuous at each point of a set D is said to be *continuous on D.* What is the largest set where the functions of the example above are continuous? The functions studied in elementary calculus are typically continuous or continuous on contiguous intervals.

Definition 1.5 *If a function f is continuous on an interval $\mathcal{I}$ except at a finite number of points, then f is called* piecewise continuous on $\mathcal{I}$.

Classifying points of discontinuity is very useful in understanding a function's behavior. We categorize these points into four kinds of discontinuity. The classifications are made on the behavior of the one-sided limits. If both exist, the discontinuity is *simple*, otherwise the discontinuity is *essential*. Subclassification of a simple discontinuity is based on whether the one-sided limits are equal. If the limit from the left equals that from the right, the discontinuity is *removable,* otherwise it is a *jump*. Subclassification of an essential discontinuity is based on whether the one-sided limits are unbounded, called an *infinite discontinuity*, or bounded, called an *oscillatory discontinuity*. These classifications are summarized in Table 1.1 and shown in Figure 1.2.

Table 1.1   Four Principal Types of Discontinuity

Simple or First Type Discontinuity

Removable:	$\lim_{x \to a} f(x) = L \neq f(a)$
Jump:	$\lim_{x \to a+} f(x) \neq \lim_{x \to a-} f(x)$

Essential or Second Type Discontinuity

Infinite:	$\lim_{x \to a+} f(x) = \pm\infty$ and/or $\lim_{x \to a-} f(x) = \pm\infty$
Oscillating:	$\lim_{x \to a+} f(x)$ and/or $\lim_{x \to a-} f(x)$ doesn't exist but is bounded.

Specific examples of each kind of discontinuity are left to the exercises.

One of the first result-oriented theorems that follows easily from the definition of continuity by using the algebra of limits describes polynomials and quotients of polynomials.

Theorem 1.5 *Polynomials are continuous everywhere. Rational functions are continuous wherever defined; the discontinuities of rational functions are either removable or infinite.*

Which class of discontinuity could contain the vertical asymptotes or poles of a rational function?

Having just used the algebra of limits and realizing that continuity is defined in terms of limits, we are led naturally to the algebra of continuity.

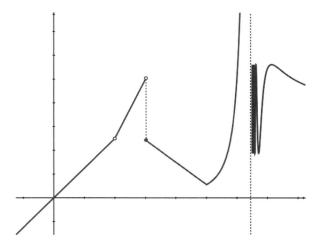

Figure 1.2   Four Types of Discontinuity

Theorem 1.6 (Algebra of Continuity) *Suppose that f and g are both continuous at $x = a$ and that $c \in \mathbb{R}$. Then*

1. *cf is continuous at a*

2. *$f \pm g$ is continuous at a*

3. *$f \cdot g$ is continuous at a*

4. *if $g(a) \neq 0$, then f/g is continuous at a*

The question of composition of functions is more subtle. The first step is to see how continuity affects limits. Since the change in the output of a continuous function is small if the change in input is small, we expect that near a point of continuity a limit would be preserved.

Theorem 1.7 *Suppose that f is continuous at $x = a$ and that $\lim_{x \to x_0} \phi(x) = a$ with the range of ϕ contained in the domain of f. Then*

$$\lim_{x \to x_0} f(\phi(x)) = f\left(\lim_{x \to x_0} \phi(x) \right)$$

Let's return to the fourth limit from Example 1.2.

■ **EXAMPLE 1.5**

- Calculate

$$\lim_{z \to 0} \sqrt[z]{1 + z}$$

Rewrite the expression as $y = (1 + z)^{(1/z)}$ and take logs of both sides to have $\ln(y) = (1/z) \cdot \ln(1 + z)$. With a little work, it can be shown that

$$z - z^2 \le \ln(1 + z) \le z$$

for $|z| < 1/2$. Divide by z, assuming that $z > 0$, and recognizing that, since $z \to 0$, we know that z is never equal to 0. Thus

$$1 - z \le \frac{\ln(1 + z)}{z} \le 1$$

The sandwich theorem tells us that $\ln(y) \to 1$ as $z \to 0+$. Similarly, $\ln(y) \to 1$ as $z \to 0$ from below. *Show it!* Hence $\lim_{z \to 0} \ln(y) = 1$. Now exponentiate this relation and apply Theorem 1.7. We arrive at

$$\lim_{z \to 0} \sqrt[z]{1 + z} = e^{\lim \ln(y)} = e^1$$

■

An immediate consequence of the "continuity preserves limits" theorem is that composition preserves continuity.

Theorem 1.8 *The composition of continuous functions is a continuous function.*

Continuous functions have many useful features. One of the most important is the *intermediate value property*: A continuous function must take on all intermediate values. This property is so geometrically natural that mathematicians such as Gauss, Euler, and Lagrange thought it obvious and used the result without proving it. Bolzano was the first to publish a rigorous argument. [See Hairer & Wanner (1996).]

Theorem 1.9 (Intermediate Value Theorem) *Let f be continuous on the closed interval $[a, b]$. If c is any value between $f(a)$ and $f(b)$, there is at least one $x \in [a, b]$ for which $f(x) = c$.*

The existence of zeros or roots is a direct application of the intermediate value theorem. If a continuous function changes sign in an interval, then it must have a zero in that interval. *Prove it!*

Another extremely valuable feature of continuous functions is that on a closed interval the function must have both a largest and a smallest value.

Theorem 1.10 (Extreme Value Theorem) *Let f be continuous on the closed interval $[a, b]$. Then f is bounded on $[a, b]$ and achieves a maximum and a minimum. That is, there are (at least) two values x_m and x_M in $[a, b]$ such that for any $x \in [a, b]$*

$$\min_{x \in [a,b]} f(x) = f(x_m) \le f(x) \le f(x_M) = \max_{x \in [a,b]} f(x)$$

The search for extreme values, maxima and minima, of a function was one of the themes that led to derivatives, our next topic. It's fascinating to realize that in the 1630s Fermat developed a method to find extreme values essentially using the derivative in the same way we do in calculus classes today—and that was before either Newton or Leibniz was born! [See Boyer (1959, pg. 155).]

1.3 DIFFERENTIATION

Just as the formal concept of limit was investigated after the mechanics of calculus were already well developed, differentiation followed long after integration as a focus of mathematical research. The derivative was originally designed to handle the "tangent problem," finding a line tangent to a curve at a point. The ancient Greeks defined a tangent as a line touching a curve only once. This definition was broadened as algebra advanced. Oresme noted in passing in the fifteenth century that "the rate of change is least ... at the point corresponding to the maximum ... " Boyer (1959, p. 85). Oresme's work was very influential and paved the way for the mathematicians of the 1600s. Fermat and others used algebraic methods for finding extrema and for mathematical modeling. In Section 1.2, we noted that Fermat's method for finding extreme values is equivalent to the process, based on derivatives, used in calculus classes today. Fermat used E in his calculations, Newton o, and Leibniz dx. The intuitive appeal of Leibniz's notation led us to use dx and Δx in current books. Newton's kinesthetic approach led to the "rate of change" motivation of modern calculus texts. Both Newton and Leibniz knew the differential triangle from Barrow's *Lectiones Opticæ et Geometricæ* (1669) shown in Figure 1.3.

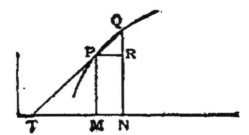

Figure 1.3   Barrow's Differential Triangle.

Typically, calculus classes study average rates of change, relating them to the slope of secant lines, and pass to the limit to obtain the instantaneous rate of change. We'll define the derivative directly using the *difference quotient* $\Delta y/\Delta x$ written in two different ways.

Definition 1.6 (Derivative)   *The* derivative *of a function f defined on an interval $\mathcal{I}$ at a point $a \in \mathcal{I}$ is given, if the limit exists, by*

$$f'(a) = \lim_{x \to a} \frac{f(x) - f(a)}{x - a}$$

The derivative function of f is given, when the limit exists, by

$$f'(x) = \lim_{\Delta x \to 0} \frac{f(x + \Delta x) - f(x)}{\Delta x}$$

If the limit exists, f is said to be differentiable *at a.*

Standard notations for the derivative of f also include $df(x)/dx$, dy/dx, and $D_x y$. Physicists often write the derivative with respect to time as $\dot{y}$; this notation is due to Newton.

The definitions of left and right hand derivatives are left to the exercises. Considering a few examples of using the limit of the difference quotient to find a derivative is our next task.

■ **EXAMPLE 1.6**

Use the definition of the derivative in the following.

1. Find $f'(0)$ when $f(x) = x^2$.
   Simple limit calculation. Do it!

2. Calculate $g'(0)$ when

$$g(x) = \begin{cases} x^3 & x \geq 0 \\ -x^3 & x < 0 \end{cases}$$

   Another simple limit, but we need to split into two cases. Do this one, too!

3. What is $U'(0)$ for the unit step function U?
   This one is just a little more subtle; slopes arbitrarily close to $a = 0$ on the left and on the right are both equal to 0. Let's look carefully.
   As x approaches 0 from above,

$$\lim_{x \to 0+} \frac{U(x) - U(0)}{x - 0} = \lim_{x \to 0+} \frac{1}{x} = \infty$$

   However, for x approaching 0 from below, we see that

$$\lim_{x \to 0-} \frac{U(x) - U(0)}{x - 0} = \lim_{x \to 0-} \frac{0}{x} = 0$$

   Hence, the limit as x approaches 0 does not exist, and therefore, the derivative $U'(0)$ does not exist.

   ■

An immediate consequence of the definition is the theorem that relates differentiability to continuity.

Theorem 1.11 *If a function f is differentiable at a, then f is continuous at a.*

The converse is not true, although Cauchy thought it was. Cauchy's was a more restrictive notion of function than ours today. Bolzano found a nondifferentiable continuous function in 1834, but his example was relatively unknown. Riemann also gave an example in his 1854 thesis. However, in 1872 Weierstrass presented an example of an everywhere continuous, nowhere differentiable function that "profoundly surprised" mathematicians. [See Kleiner (1989).] Today, the absolute value function is the canonical example in current calculus texts showing that continuity at a point does not give differentiability there.

Just as with limits and with continuity, the algebra of derivatives extends our toolbox. The proofs in the following result are relatively easy applications of the definition, appear in many elementary calculus textbooks, and here are left to the reader as an exercise.

Theorem 1.12 (Algebra of Derivatives) *Suppose that f and g are differentiable at* $x = a$ *and that* $c \in \mathbb{R}$. *Then*

1. $(c \cdot f)'(a) = c \cdot f'(a)$

2. $(f \pm g)'(a) = f'(a) \pm g'(a)$

3. $(f \cdot g)'(a) = f'(a) \cdot g(a) + f(a) \cdot g'(a)$

4. *if* $g(a) \neq 0$, *then* $\left(\dfrac{f}{g}\right)'(a) = \dfrac{f'(a) \cdot g(a) - f(a) \cdot g'(a)}{g^2(a)}$

With these techniques in hand, we can easily differentiate any polynomial or rational function. For transcendental functions, we use the limit definition of the derivative or trigonometric identities to develop formulas. The three basic results follow.

Theorem 1.13 *For any real number* x

1. $\dfrac{d}{dx} \sin(x) = \cos(x)$

2. $\dfrac{d}{dx} e^x = e^x$

3. $\dfrac{d}{dx} \ln(x) = \dfrac{1}{x}$ *for* $x > 0$

Many texts first establish the *generalized power rule*

$$\frac{d}{dx} f^r(x) = r f^{r-1}(x) f'(x)$$

for $r \in \mathbb{Q}$ as part of the development of the *chain rule*. Since the power rule is a direct corollary, we'll proceed directly to the chain rule.

Theorem 1.14 (Chain Rule) *Let* $f : [a, b] \to [c, d]$ *and* $g : [c, d] \to \mathbb{R}$ *be differentiable functions at* $x = a$ *and* $x = f(a)$, *respectively. Then* $g \circ f : [a, b] \to \mathbb{R}$ *is differentiable at* $x = a$ *and its derivative is given by*

$$(g \circ f)'(a) = g'(f(a)) \cdot f'(a)$$

The name of the result comes from the form

$$\frac{dy}{dt} = \frac{dy}{dx} \cdot \frac{dx}{dt}$$

appearing to chain derivatives together.

The chain rule is an extremely powerful tool but has a nontrivial proof which causes most elementary calculus texts to focus on motivation, heuristics, and usage. We'll defer the proof until later. One of the nicer "mathematical applications" of the chain rule is to develop a formula for the derivative of an inverse function. Since $f \circ f^{-1}$ is the identity function, we see that

$$(f \circ f^{-1})'(x) = 1$$

Applying the chain rule tells us that

$$f'(f^{-1}(x)) \cdot f^{-1'}(x) = 1$$

Employing just a little algebra, assuming proper assumptions preventing dividing by zero, yields a general result.

Theorem 1.15 *If $f : [a, b] \to \mathbb{R}$ is differentiable and $f^{-1}(x) \neq 0$ on $[a, b]$, then, for $y = f(x)$,*

$$f^{-1'}(y) = \frac{1}{f'(x)}$$

Now that we have the basic mechanics of differentiation in hand, there are two directions to go. We can study the geometry of derivatives and higher order derivatives, and we can investigate applications. First, let's consider geometry.

Since the derivative describes the instantaneous rate of change of a function, it tells us whether a function is momentarily increasing, decreasing, or constant. Values where the derivative is zero, where the function has a horizontal tangent, are called *critical values* or *stationary points*. [Note: some calculus authors, e.g., Stewart (2009), add values where the derivative fails to exist to the definition of critical value; others, e.g., Thomas (1968), also add endpoints of the domain.] A stationary point where a function does not change from increasing to decreasing or vice versa is called a *terrace point*. We can further analyze the change in the derivative, the change in that change, etc. Let's define higher order derivatives recursively.

Definition 1.7 *Let f be a differentiable function and $n \in \mathbb{N}$. Then for $n > 1$ the nth derivative of f, denoted by $f^{(n)}(x)$, is given, when it exists, by*

$$f^{(n)}(x) = \frac{d}{dx} f^{(n-1)}(x) \qquad or \qquad \frac{d^n y}{dx^n} = \frac{d}{dx} \frac{d^{n-1} y}{dx^{n-1}}$$

Some authors (and calculators like the TI-Nspire) adopt the convention that the zeroth derivative is the original function; i.e., $f^{(0)}(x) = f(x)$. *Rewrite the definition above using this convention.* A little classroom fun can be had by asking calculus students, "What is the 'leap year'th derivative of the sine?"

Geometrically interpreting an increasing rate of change as a graph curving up, we call it *concave up*. A decreasing rate of change is called *concave down*. Our definitions follow Granville, Smith, and Longley (1911) and use "positive" versus "nonnegative," etc., to eliminate linear functions being simultaneously both concave up and down.

Definition 1.8 *Let f be a twice-differentiable function on the interval $\mathcal{I}$. Then:*

- *If $f''(x)$ is positive on $\mathcal{I}$, then $f'(x)$ is increasing and f is* concave up *on $\mathcal{I}$.*

- *If $f''(x)$ is negative on $\mathcal{I}$, then $f'(x)$ is decreasing and f is* concave down *on $\mathcal{I}$.*

- *If f changes concavity about a point x_0, then x_0 is an* inflection point *of f.*

In economics, an inflection point of an increasing function changing from concave up to down is often called a *point of diminishing return* since a corresponding change in input produces a smaller change in output as x passes the inflection point. *Explain why this happens.*

Write this geometric information algebraically to define procedures equivalent to Fermat's method, but more general, for finding local extreme values of a function.

Theorem 1.16 (First Derivative Test for Extrema) *Suppose that f is differentiable on an interval containing the critical value a.*

- *If f' changes from positive below a to negative above a, then $f(a)$ is a local maximum of f.*

- *If f' changes from negative below a to positive above a, then $f(a)$ is a local minimum of f.*

- *If f' does not change sign about a, then $(a, f(a))$ is a* terrace point *and not a local extreme value.*

Using the information on concavity provided by the second derivative yields a simpler test for local extrema.

Theorem 1.17 (Second Derivative Test for Extrema) *Suppose that f is twice differentiable on an interval containing the critical value a.*

- *If $f''(a) < 0$, then $f(a)$ is a local maximum of f.*

- *If $f''(a) > 0$, then $f(a)$ is a local minimum of f.*

- *If $f''(a) = 0$, then the test fails.*

Let's use the famous *Norman window problem* as an illustration.

■ **EXAMPLE 1.7   The Norman Window Problem**

A Norman window is a window in the shape of a rectangle surmounted by a semicircle. See Figure 1.4. If the perimeter of the window is P, find the dimensions allowing the greatest amount of light to be admitted.

Figure 1.4   A Norman Window

Solution. Let the rectangle have height h and width d. Then the diameter of the semicircle is also d. Thus, the perimeter is

$$P = (2h + d) + \frac{\pi}{2} d$$

Since the light admitted is directly proportional to the area A, we need to maximize A. That is, we need to maximize

$$A = hd + \frac{\pi}{8} d^2$$

Solve the perimeter equation for h and substitute the result into the area equation:

$$A = \frac{P}{2} d - \left(\frac{\pi}{8} + \frac{1}{2} \right) d^2$$

Set the derivative of A with respect to d equal to 0 :

$$A' = \frac{P}{2} - \left(\frac{\pi}{4} + 1 \right) d = 0$$

The critical value is seen to be

$$d_c = \frac{2P}{\pi + 4}$$

Since $A'' = -1 - \pi/4 < 0$, we must have a maximum when the diameter and width are $2P/(\pi + 4)$ and the height is $P/(\pi + 4)$. ∎

The Norman window problem has been a favorite of calculus authors for many, many years. See, for example, Granville et al. (1911, p. 57), Thomas (1968, p. 129),

Ellis et al. (1999, p. 189), Hughes-Hallett et al. (2009, p. 263), Stewart (2009, p. 208) (tunnel), etc.

Continuing with our geometric theme, we consider what happens to a differentiable function between points having equal functional values. If the function rises, then it must smoothly turn and come back down. If the function sinks, it must smoothly turn and come back up. A function that neither increases nor decreases must be constant. Whatever the case, there must be at least one point with a horizontal tangent on the graph. This result is known as Rolle's theorem.

Theorem 1.18 (Rolle's Theorem) *Let f be a function that is continuous on $[a, b]$ and differentiable on (a, b). If $f(a) = f(b)$, then there is a value $c \in (a, b)$ such that $f'(c) = 0$.*

The natural generalization of Rolle's theorem is to let the endpoints be at different levels. The generalized theorem was first stated by Lagrange.

Theorem 1.19 (Mean Value Theorem) *Let f be a function that is continuous on $[a, b]$ and differentiable on (a, b). Then there is a value $c \in (a, b)$ such that*

$$f'(c) = \frac{f(b) - f(a)}{b - a}$$

Both Rolle's theorem and Lagrange's mean value theorem are *existence theorems*; that is, these theorems specify that there is some value satisfying the conclusion but give no direction on how to find that value. Nevertheless, existence theorems often have very practical ramifications. Two important corollaries of the mean value theorem concern properties of differentiable functions.

Corollary 1.20 *If $f'(x) = 0$ for all $x \in (a, b)$, then f is a constant function on (a, b).*

The unit step function has a zero derivative everywhere except at $a = 0$ yet is not a constant function. How does this not violate the corollary? Now apply this idea to two functions.

Corollary 1.21 *If $f'(x) = g'(x)$ for all $x \in (a, b)$, then $f(x) = g(x) + c$ for all $x \in (a, b)$ for some constant c.*

The last application of the mean value theorem we will consider is known as the *speed limit law;* cf. Ostebee & Zorn (2002, p. 42). This law is the first step on the road to Taylor series, as we'll see later in Section 1.6.

Theorem 1.22 (Speed Limit Law) *Suppose that f is differentiable on (a, b) and continuous on $[a, b]$ and that*

$$m \leq f'(x) \leq M$$

for all $x \in (a, b)$ for two constants m and M. Then

$$m(b - a) \leq f(b) - f(a) \leq M(b - a)$$

Further, if $x \in [a, b]$, then

$$f(a) + m(x - a) \leq f(x) \leq f(a) + M(b - a)$$

The speed limit law defines a cone containing the function as shown in Figure 1.5. We are able to find upper and lower bounds for the values of f between a and b.

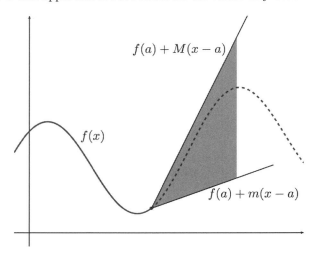

$f(a) + M(x - a)$

$f(x)$

$f(a) + m(x - a)$

Figure 1.5   The "Speed Limit Cone"

A two-function variant of the speed limit law, named by Jerry Uhl, Professor Emeritus of Mathematics at the University of Illinois, is known as the *race track principle* (Davis et al., 1994).

Theorem 1.23 (Race Track Principle) *Suppose that f and g are continuous on $[a, b]$ and differentiable on (a, b) and that $f'(x) \leq g'(x)$ for $a < x < b$:*

1. *If $f(a) = g(a)$, then $f(x) \leq g(x)$ for $a \leq x \leq b$.*

2. *If $f(b) = g(b)$, then $f(x) \geq g(x)$ for $a \leq x \leq b$.*

The name comes from interpreting the derivative as velocity. The principle then says that if two horses start a race together, the faster horse is always in the lead. The second statement says that if two horses finish together, the faster horse was always behind.

Derivatives are incredibly useful in real life. Since change is much easier to measure than quantity, mathematical models of continuous phenomena are often specified as differential equations. The simplest models have the form $y' = f(x)$. The exponential growth of Malthusian population dynamics comes from the model $p' = r \cdot p$, where p represents a population at time t and r is a constant. These models ask the question, "What function has the specified derivative?" The connection between differentiation and integration then becomes the focus of our study.

1.4 INTEGRATION

Integration is the oldest part of calculus and developed from the need for formulas for areas and volumes of geometric figures. From Hippocrates of Chios' "Quadrature of the Lune" [see Dunham (1990)] onward, the search to understand area and volume has led to deeper mathematical discoveries. The Greek mathematician Eudoxus of Cnidus' *method of exhaustion* (about 370 B.C.) is the progenitor of the definite integral as a limit of approximating sums. The method of exhaustion was *the* technique used to calculate area and volume formulas for 2000 years. The word *quadrature* originally meant constructing, with only a straightedge and compass, a square having exactly the same area as a given figure but later came to mean determining an area or volume. Modern usage of the term denotes "numerical integration."

A number of basic results for definite integrals had been determined before either Newton or Leibniz studied mathematics. For example, the formula

$$\int_0^a x^n dx = \frac{a^{n+1}}{n+1},$$

appears (in very different notation) in the work of Cavalieri (1635), Torricelli (1635), Roberval (1636), Fermat [c. 1644, Boyer (1945)], Pascal [1654, Boyer (1959, p. 149)], and Wallis [1665, Burton (2007, p. 386)]. It was the genius of Newton and Leibniz that they saw the significance of the connection between integration and differentiation, that these are inverse operations—the *fundamental theorem of calculus*. Isaac Barrow, Newton's teacher and predecessor in the Lucasian Chair at Cambridge, had formulated the fundamental theorem (published in his 1670 monograph *Lectiones Geometricæ*, or *Lectures on Geometry*, after he had left Cambridge) but had not understood the theorem's importance since his focus was very geometrical, not analytic.

Most calculus texts motivate the definite integral via calculating the area enclosed between a curve and the x-axis. We'll begin our look at integration by directly defining the definite integral as a limit and then move to the fundamental theorem and applications. However, texts diverge in their definitions. Some use equipartitions [e.g., Stewart (2009)], while others use more general subdivisions [e.g., Thomas (1968)]. Since we'll be studying the integral in much more generality later, we'll begin with Riemann sums. First, we define a *partition* and *sample points* or *tags*.

Definition 1.9 *A* partition *of the interval* $[a, b]$ *is a set of* $n + 1$ *values*

$$\mathcal{P} = \{a = x_0 < x_1 < x_2 < \cdots < x_{n-1} < x_n = b\}$$

that divides $[a, b]$ *into* n *subintervals with widths* $\Delta x_k = x_k - x_{k-1}$. *Sample points or tags are values* c_k *chosen in each subinterval of* $\mathcal{P}$; *that is, for each* $k = 1, \ldots, n$, *there is a* $c_k \in [x_{k-1}, x_k]$. *A partition together with a set of tags will be denoted by* $\mathcal{P}_c$.

Now that we've partitioned the interval and chosen sample points, we can build a Riemann sum.

Definition 1.10 (Riemann Sum) *Let f be a function defined on $[a, b]$. The* Riemann sum *of f over $\mathcal{P}_c$, the partition and chosen sample points, is*

$$\mathcal{R}_{\mathcal{P}_c}(f) = \sum_{k=1}^{n} f(c_k)\Delta x_k$$

Passing to the limit as the largest Δx_k goes to 0 gives us the definite integral of f over $[a, b]$. Set $\|\mathcal{P}_c\| = \max_k \Delta x_k$.

Definition 1.11 (Riemann Integral) *Let f be a function defined on $[a, b]$. The* Riemann integral *of f over $[a, b]$ is, if the limit exists,*

$$\int_a^b f(x)\, dx = \lim_{\|P_c\| \to 0} \sum_{k=1}^{n} f(c_k)\Delta x_k$$

There are subtleties to this limit that many elementary calculus texts finesse by using only partitions with uniform widths so that $\Delta x_k = \Delta x = (b - a)/n$. Example calculations are also much simpler with uniform partitions as the limit reduces to

$$\int_a^b f(x)\, dx = \lim_{n \to \infty} \sum_{k=1}^{n} f(c_k)\Delta x$$

as shown in Figure 1.6.

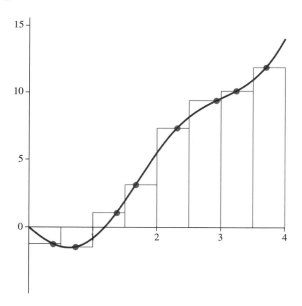

Figure 1.6   A Uniform Width Riemann Partition

A typical example follows.

■ **EXAMPLE 1.8**

Calculate $\int_0^1 x^3 \, dx$.

Choose the partition $\{x_k = k/n \mid k = 0, \ldots, n\}$ with the sample points $c_k = k/n$ for $k = 1, \ldots, n$. Then $\Delta x = 1/n$, and thus

$$\int_0^1 x^3 \, dx = \lim_{n \to \infty} \sum_{k=1}^{n} \left(\frac{k}{n}\right)^3 \cdot \frac{1}{n}$$

$$= \lim_{n \to \infty} \frac{1}{n^4} \cdot \sum_{k=1}^{n} k^3$$

$$= \lim_{n \to \infty} \frac{1}{n^4} \cdot \frac{n^2(n+1)^2}{4}$$

$$= \frac{1}{4}$$

■

Just as we did for limits, continuity, and differentiation, we extend our toolbox by showing that integration is a linear operation.

Theorem 1.24 (Algebra of Integrals) *Suppose that f and g are integrable on* $[a, b]$ *and that* $c \in \mathbb{R}$. *Then*

1. $\displaystyle\int_a^b c \cdot f(x) \, dx = c \cdot \int_a^b f(x) \, dx$

2. $\displaystyle\int_a^b (f \pm g)(x) \, dx = \int_a^b f(x) \, dx \pm \int_a^b g(x) \, dx$

The proofs are easy verifications using the definition.

An image of a function inscribed in one rectangle and circumscribing another gives a "proof by picture" for our next result. *Draw the image!* The proof is left to the exercises.

Theorem 1.25 *Let f be integrable on* $[a, b]$. *If there are two constants m and M such that* $m \le f(x) \le M$ *for all* $x \in [a, b]$, *then*

$$m \cdot (b - a) \le \int_a^b f(x) \, dx \le M \cdot (b - a)$$

A third rectangle can be drawn with the rectangle's area equal to the function's. This rectangle's height gives the *average value* of the function.

Definition 1.12 (Average Value of a Function) *If f is integrable on* $[a, b]$, *then the average value of f is given by*

$$\bar{f} = \frac{1}{b - a} \cdot \int_a^b f(x) \, dx$$

■ EXAMPLE 1.9

The average value depends on the interval. The average value of the sine function over $[0, 2\pi]$ is

$$\overline{\sin} = \frac{1}{2\pi} \cdot \int_0^{2\pi} \sin(x)\, dx = 0$$

while the average over $[0, \pi]$ is

$$\overline{\sin} = \frac{1}{\pi} \cdot \int_0^{\pi} \sin(x)\, dx = \frac{2}{\pi} \approx 0.63662$$

If f is continuous, then, just as with derivatives, the function takes on its average value. This result is proven by showing that the function must intersect that middle rectangle.

Theorem 1.26 (Mean Value Theorem for Integrals) *If f is continuous on $[a, b]$, then there is at least one point $c \in (a, b)$ where $f(c) = \overline{f}$.*

At this point it is natural to ask, "What type of function has a definite integral?" An easy to prove answer is continuous.

Theorem 1.27 *If f is continuous on $[a, b]$, then $\int_a^b f(x)\, dx$ exists.*

In 1823, Cauchy used uniform continuity to show that given $\epsilon > 0$ there is a $\delta > 0$ so that any two partitions with $\max_k \Delta x_k < \delta$ have to have Riemann sums within ϵ of each other. This argument proved the limit existed without finding the value of the limit. We will use this method again later when we consider sequences in Section 1.5. Cauchy's technique is based strongly on the structure of the real numbers; it does not hold if we only allow ourselves to use rational numbers.

The title *fundamental theorem* indicates that this result forms the core of calculus, linking the two main concepts, derivative and integral. While special cases were known previously, Newton and Leibniz were the first to realize that the theorem provided a new, general form of analysis connecting the tangent problem to quadrature and vice versa. The theorem has two statements: The integral is differentiable and the antiderivative gives the quadrature.

Theorem 1.28 (Fundamental Theorem of Calculus) *Let f be a continuous function on the interval $[a, b]$.*

1. *Define $F(x) = \int_a^x f(x)\, dx$. Then F is continuous and differentiable on $[a, b]$ with $F'(x) = f(x)$.*

2. *If F is any antiderivative of f, then $\int_a^b f(x)\, dx = F(b) - F(a)$.*

Boyer (1959, p. 280) tells us that Cauchy was the first to give a rigorous proof of the fundamental theorem.

The next three results form the basis of a good portion of the second half of an elementary calculus course: substitution, integration by parts, and partial fraction integration. All three address the question of how to find an antiderivative; all three are reductions to simpler forms, not results. Substitution is the inverse of the chain rule.

Theorem 1.29 (Integration by Substitution) *Let f be a continuous function of u and u be a continuously differentiable function of x on $[a, b]$. Then*

$$\int f(u)\,du = \int f(u(x))u'(x)\,dx$$

Integration by parts is the inverse of the product rule.

Theorem 1.30 (Integration by Parts) *Let f and g be continuously differentiable functions on $[a, b]$. Then*

$$\int_a^b f(x)g'(x)\,dx = f(x)g(x)\Big|_a^b - \int_a^b g(x)f'(x)\,dx$$

This rule is often written with differentials in the form $\int u\,dv = uv - \int v\,du$.

The last of the three rules is really an algebraic technique for rational functions. Any rational function $R(x)$ can be decomposed into a sum of a polynomial p and terms having the form $a/(x - b)^k$ where b is a root of the denominator and a and k are constants. The partial fraction expansion gives the form

$$R(x) = p(x) + \sum_{k=1}^n \left(\frac{a_{k1}}{x - b_k} + \frac{a_{k2}}{(x - b_k)^2} + \cdots + \frac{a_{km_k}}{(x - b_k)^{m_k}} \right)$$

for a rational function $R(x)$ with denominator roots b_k having multiplicity m_k. *Integration by partial fractions* is just applying the integral to both sides of the equation above. Beginning with his 1833 paper on integrals that are algebraic, which heavily used partial fractions, Liouville analyzed the process of integration to find which functions could or could not have antiderivatives with finite elementary forms. [See Kasper (1980) for a nice discussion of Liouville's work and where it led others.] Liouville's approach was very influential in the development of computer algebra systems. We now can use software like Maple and Mathematica and calculators such as Texas Instruments' Voyage 200, TI-89, and TI-Nspire CAS to calculate integrals that would be very difficult by hand.

Calculus texts usually include tables having over a hundred formulas for antiderivatives. The *CRC Standard Mathematical Tables and Formulae* (Zwillinger, 2002) contains over 600 antiderivative forms, Gradshteyn and Ryzhik's *Table of Integrals, Series, and Products* (Gradshteyn & Ryzhik, 2007) countains thousands. Even though techniques of integration form a major focus of second-semester calculus and there are texts like the *Handbook of Integration* (Zwillinger, 1992), most functions do not have elementary antiderivatives. Extremely powerful computers and calculators have become ubiquitous, leading to a resurgence of approximation theory. These observations provide motivation for us to study approximate methods of integration called *numerical quadratures*. We will consider three of the standard techniques that are used in elementary courses. The methods use constant, linear, and quadratic approximations, respectively.

For the constant approximation, we will use the left endpoint to measure the function's height, effectively giving us Cauchy's definition of the definite integral; the associated error is proportional to the first derivative. Other choices of where to measure the height include the midpoint and right endpoint. The trapezoid rule uses the linear approximation through the left and right endpoints with an error proportional to the second derivative. Simpson's rule uses quadratic approximation over the endpoints of two consecutive intervals; the error bound is proportional to the fourth derivative.

Theorem 1.31 (Numerical Integration Methods) *Suppose that* $\mathcal{I} = \int_a^b f(x)\,dx$ *exists. Given* $n \in \mathbb{N}$, *set* $\Delta x = (b-a)/n$ *and* $x_k = a + k\Delta x$ *with* $k = 0, \ldots, n$. *Also, let* M_k *be an upper bound for* $|f^{(k)}(x)|$ *on* $[a, b]$. *Then:*

Left Sum *The left sum approximation* L_n *to* $\mathcal{I}$ *with error bound* ε_n *is*

$$L_n = \sum_{k=1}^{n} f(x_{k-1})\Delta x \qquad \varepsilon_n = |L_n - \mathcal{I}| \leq M_1 \cdot \frac{(b-a)^2}{2} \cdot \frac{1}{n}$$

Trapeziod Rule *The trapezoid rule approximation* T_n *to* $\mathcal{I}$ *with error bound* ε_n *is*

$$T_n = \sum_{k=1}^{n} \frac{1}{2}\left(f(x_{k-1}) + f(x_k)\right)\Delta x \qquad \varepsilon_n = |T_n - \mathcal{I}| \leq M_2 \cdot \frac{(b-a)^3}{12} \cdot \frac{1}{n^2}$$

Simpson's Rule *The Simpson's rule approximation* S_n *(n must be even) to* $\mathcal{I}$ *with error bound* ε_n *is*

$$S_n = \sum_{k=1}^{n/2} \frac{1}{3}\left(f(x_{2k-2}) + 4f(x_{2k-1}) + f(x_{2k})\right)\Delta x$$

$$\varepsilon_n = |S_n - \mathcal{I}| \leq M_4 \cdot \frac{(b-a)^5}{180} \cdot \frac{1}{n^4}$$

In calculus classes, bounds for M_k are usually found graphically. Table 1.2 summarizes the results of applying the quadrature rules to $\int_0^3 e^{-x^2}\,dx$.

Finding an antiderivative is much more challenging than calculating a derivative. It's not easy to know a priori whether or not a given function has an elementary antiderivative. Sometimes it's easy to prove as with $f(x) = e^{-x^2}$. Other times, it's very hard to determine if an elementary antiderivative exists. This difficulty makes integration quite a challenge. One of the questions we ask is if the values of the Riemann sums from the definition of the definite integral converge. This question leads to our next topic, sequences and series of constants.

Table 1.2 Numerical Integration of $\int_0^3 e^{-x^2}\, dx$

n	L_n	ε_n	T_n	ε_n	S_n	ε_n
4	1.26113	0.37493	0.886180	0.000027	0.886206	$1.7297 \cdot 10^{-6}$
10	1.07368	0.18747	0.886199	0.000008	0.886207	$1.398 \cdot 10^{-7}$
20	0.96120	0.07499	0.886206	0.000001	0.886207	$4.0 \cdot 10^{-9}$

1.5 SEQUENCES AND SERIES OF CONSTANTS

Some students argue that $1 \neq 0.\bar{9}$. They are unwittingly reprising Zeno's paradox from the fifth century B.C. The most familiar version of the paradox states that Achilles can never catch a tortoise as first he must cover half the distance to the tortoise, then half the remaining distance, then again half the remaining distance, ad infinitum. What Zeno objected to was infinite divisibility. We overcome the problem with the concept of *convergence*, basing it on our formal definition of limit. In this section, we will consider sequences and series of constants and allow Achilles ultimately to catch the tortoise.

Sequences

A real-valued sequence is a real-valued function that has a special domain.

Definition 1.13 *A* sequence *is a function from the natural numbers* $\mathbb{N}$ *to the real numbers. For a sequence* $a : \mathbb{N} \to \mathbb{R}$, *we denote* $a(n)$ *by* a_n *and the sequence by* $\{a_n\}$.

Zeno's paradox concerns the sequence $\{1/2, 1/4, 1/8, \dots\}$. The paradox arises from the fact that every term is greater than zero, but we know that the terms eventually get smaller than any given positive number. We need to define the limit of a sequence in order to let Achilles catch the tortoise.

Definition 1.14 *Let* $\{a_n\}$ *be a sequence. Then* L *is the limit of* $\{a_n\}$ *or*

$$\lim_{n \to \infty} a_n = L$$

if and only if for every $\epsilon > 0$ *there is an* $N \in \mathbb{N}$ *such that whenever* $n > N$ *it must follow that*

$$|a_n - L| < \epsilon$$

If $\lim_{n \to \infty} a_n = L$, then we say the sequence *converges to* L; a sequence that does not converge is said to *diverge*. It's not hard to prove that Zeno's sequence converges to zero. *Do it!*

In 1202, Leonardo of Pisa's book *Liber Abaci* introduced Hindu-Arabic numbers and positional notation to Europe. However, the text is better known for a famous sequence arising from an example:

> A certain man put a pair of rabbits in a place surrounded on all sides by a wall. How many pairs of rabbits can be produced from that pair in a year if it is

supposed that every month each pair begets a new pair which from the second month on becomes productive?

The solution to the problem is the sequence $\{1, 1, 2, 3, 5, 8, \dots\}$, which is now called, using Leonardo's nickname, the *Fibonacci sequence*. We show the Fibonacci sequence diverges by combining two facts: first, all terms are positive; second, each term beyond the second is larger than the previous by at least one.

Let's look at two sequences that may not obviously converge or diverge at first glance.

■ **EXAMPLE 1.10**

1. Does the sequence with general term $a_n = (\sqrt{n^2 + n} - n)/n$ converge or diverge?

   First, rationalize the numerator:

   $$a_n = \frac{\sqrt{n^2 + n} - n}{n} \cdot \frac{\sqrt{n^2 + n} + n}{\sqrt{n^2 + n} + n} = \frac{1}{\sqrt{n^2 + n} + n}$$

   Now, it is easy to see that a_n converges to 0.

2. Does the sequence with general term $b_n = (c + 1)(-1)^n - 1$ converge or diverge for an arbitrary $c \in \mathbb{R}$?

   Write several terms. What do you see? (Be careful—there are two cases.) ■

Once again, it's time for an algebra, now of sequences. The proofs are all straightforward, essentially restatements of earlier limit theorems for functions, and are left to the exercises.

Theorem 1.32 (Algebra of Sequence Limits) *Let $\{a_n\}$ and $\{b_n\}$ be convergent sequences and let $c \in \mathbb{R}$. Then*

1. $\displaystyle\lim_{n \to \infty} (c \cdot a_n) = c \cdot \lim_{n \to \infty} a_n$

2. $\displaystyle\lim_{n \to \infty} (a_n \pm b_m) = \lim_{n \to \infty} a_n + \lim_{n \to \infty} b_n$

3. $\displaystyle\lim_{n \to \infty} (a_n \cdot b_m) = \lim_{n \to \infty} a_n \cdot \lim_{n \to \infty} b_n$

4. $\displaystyle\lim_{n \to \infty} (a_n/b_m) = \lim_{n \to \infty} a_n / \lim_{n \to \infty} b_n$ *provided* $\lim b_n \neq 0$

The sandwich theorem can also be restated for sequences.

Theorem 1.33 (The Sandwich Theorem for Sequences) *Let $\{a_n\}, \{b_n\},$ and $\{c_n\}$ be sequences for which there is an $N \in \mathbb{N}$ such that for all $n > N$ we have $a_n \leq b_n \leq c_n$. If $\lim_{n \to \infty} a_n = L$ and $\lim_{n \to \infty} c_n = L$, then $\lim_{n \to \infty} b_n = L$.*

The sandwich theorem limits the growth of a sequence. At times, it is useful to see if a sequence can describe its own action. We can classify sequences as increasing or decreasing.

Definition 1.15 *A sequence* $\{a_n\}$ *is*

- increasing *if* $a_n < a_{n+1}$ *for all* n,

- decreasing *if* $a_n > a_{n+1}$ *for all* n,

- bounded *if* $|a_n| \leq M$ *for all* n *for some real number* M.

A monotone sequence is either increasing or decreasing.

The properties of bounded and monotone are very strong and lead to important results, such as the following due to Bolzano and Weierstrass.

Theorem 1.34 (Bolzano-Weierstrass Theorem for Sequences) *If* $\{a_n\}$ *is a bounded sequence of real numbers, then* $\{a_n\}$ *has a subsequence that converges.*

Bolzano discovered the theorem first, but his work was relatively unknown; some of his manuscripts were not even published until the 1960s (Burton, 2007, p. 671). Weierstrass independently proved the result in the 1860s (Burton, 2007, p. 688). An immediate corollary of the theorem is that *a bounded, monotone sequence converges.*

Series

A *series* $\{s_n\}$ is a special sequence in which the elements are formed from a sequence $\{a_n\}$ by

$$s_n = \sum_{k=1}^{n} a_k.$$

Each s_n is called a *partial sum*. We often write $\sum a_n$ to represent the series. A series converges if its sequence of partial sums converges. We can further classify convergence as follows.

Definition 1.16 (Absolute and Conditional Convergence) *A series* $\sum a_n$ *is absolutely convergent if the series* $\sum |a_n|$ *converges. If* $\sum a_n$ *converges but* $\sum |a_n|$ *diverges, then* $\sum a_n$ *is called* conditionally convergent.

The series $\sum (-1/2)^k$ is absolutely convergent while $\sum (-1)^k/k$ is conditionally convergent. We'll develop tests to verify these statements in a moment.

Zeno's claim rephrased in terms of series is

$$\sum_{k=1}^{n} \frac{1}{2^k} < 1 \quad \text{for all } n \in \mathbb{N}$$

But what happens to Zeno's series as n goes to infinity? Does the series converge? Before looking at different tests for convergence, let's consider examples of convergent and divergent series. Explore these series numerically and graphically using different values for their parameters. *Zeno's series belongs to which type below?*

■ **EXAMPLE 1.11   Special Series of Elementary Calculus**

1. A *geometric series* $\sum ar^k$ converges when $|r| < 1$ with

$$\sum_{k=0}^{\infty} a\, r^k = \frac{a}{1 - r}$$

2. The *harmonic series* $\sum 1/k$ diverges since

$$\sum_{k=1}^{n} \frac{1}{k} \approx \ln(n)$$

3. A *p-series* $\sum 1/k^p$ converges if $p > 1$ and diverges for $p \leq 1$ since

$$\sum_{k=1}^{n} \frac{1}{k^p} \approx \int_{1}^{n} \frac{1}{x^p}\, dx = \frac{n^{1-p} - 1}{1 - p}$$

4. The *telescoping series* $\sum 1/(k(k+1))$ and $\sum 1/((2k-1)(2k+1))$ converge to 1 and $1/2$, respectively, since

$$\sum_{k=1}^{n} \frac{1}{k(k+1)} = \sum_{k=1}^{n} \left(\frac{1}{k} - \frac{1}{k+1} \right) = 1 - \frac{1}{n+1}$$

and

$$\sum_{k=1}^{n} \frac{1}{(2k-1)(2k+1)} = \sum_{k=1}^{n} \frac{1}{2} \left(\frac{1}{2k-1} - \frac{1}{2k+1} \right) = \frac{1}{2} \left(1 - \frac{1}{2n+1} \right)$$

■

Oresme showed, in the fourteenth century, that the harmonic series diverges with a technique that is still used in classes today (Struik, 1986, p. 320). Collect the terms of the series as

$$1 + \left(\frac{1}{2} \right) + \left(\frac{1}{3} + \frac{1}{4} \right) + \left(\frac{1}{5} + \frac{1}{6} + \frac{1}{7} + \frac{1}{8} \right) + \cdots$$
$$> 1 + \left(\frac{1}{2} \right) + \left(\frac{1}{4} + \frac{1}{4} \right) + \left(\frac{1}{8} + \frac{1}{8} + \frac{1}{8} + \frac{1}{8} \right) + \cdots$$
$$= 1 + \left(\frac{1}{2} \right) + \left(\frac{1}{2} \right) + \left(\frac{1}{2} \right) + \cdots$$

taking groups of size 2^n. Since the last series is growing infinitely large, the harmonic series, being larger, must diverge.

If a series we are investigating isn't one of the standards, how do we determine convergence? If the general term does not go to zero, then the series must diverge. What about positive answers? We need a collection of criteria. The easiest procedure to apply is the ratio test.

Theorem 1.35 (D'Alembert's Ratio Test) *Let $\sum a_n$ be a series with positive terms. Set*

$$r = \lim_{n \to \infty} \frac{a_{n+1}}{a_n}.$$

Then:

- *If $r < 1$, the series converges.*

- *If $r > 1$, the series diverges.*

- *If $r = 1$, the test fails.*

The proof of the ratio test is an application of geometric series convergence. While the ratio test is usually easy to apply, it doesn't always prove conclusive. Try the ratio test on $\sum 1/n^3$. The root test is more sensitive but can be harder to use.

Theorem 1.36 (Cauchy's Root Test) *Let $\sum a_n$ be a series with positive terms. Set*

$$\rho = \lim_{n \to \infty} \sqrt[n]{a_n}$$

Then:

- *If $\rho < 1$, the series converges.*

- *If $\rho > 1$, the series diverges.*

- *If $\rho = 1$, the test fails.*

The comparison test is easier to use than the root test. The difficulty is that we must use a series with known convergence or divergence to compare to the series we're analyzing.

Theorem 1.37 (Comparison Test) *Let $\sum a_n$ and $\sum b_n$ be series of positive terms with $a_n \leq b_n$ for all n.*

- *If $\sum b_n$ converges, then so does $\sum a_n$.*

- *If $\sum a_n$ diverges, then so does $\sum b_n$.*

Another sort of comparison test is based on Riemann sums. If we partition $[1, \infty)$ with the natural numbers, then $\Delta x = 1$ and the Riemann sum becomes the summation we're considering. Hence, the integral and the sum either both converge or both diverge.

Theorem 1.38 (Integral Test) *Suppose that f is continuous, positive, and decreasing for $x \geq 1$ and that $a_n = f(n)$ for all n.*

- *If $\displaystyle\int_1^\infty f(x)\,dx$ converges, then so does $\sum a_n$.*

- *If $\displaystyle\int_1^\infty f(x)\,dx$ diverges, then so does $\sum a_n$.*

The last test we consider, due to Leibniz, is for a special type of series. An *alternating series* has terms that alternate in sign and monotonically go to zero.

Theorem 1.39 (Alternating Series Test) *Let $\{a_n\}$ be a positive, decreasing sequence that converges to zero. Then the alternating series*

$$\sum_{n=1}^\infty (-1)^{n-1} a_n$$

converges.

It's interesting to note that the *tail* of an alternating series $\sum_{k=n+1}^\infty (-1)^{n-1} a_k$ is bounded in absolute value by a_{n+1}.

Each of the convergence tests is an analog of an existence theorem, not providing a limit value, just indicating whether or not a series converges. For more delicate tests, see, for example, Wrede & Spiegel (2002) or, for detailed explanations, see Kosmala (2004).

We have focused on sequences and series of constants in this section, getting ready to study sequences and series of functions. Part of Newton's success was his facility with expressing functions as series and manipulating them with his new methods of calculus. As Newton knew and we'll see, series of functions are a powerful tool.

1.6   POWER SERIES AND TAYLOR SERIES

Geometric series had been widely used before the calculus methods of Newton and Leibniz were invented. Nicolas Mercator used a geometric series to develop a power series representation for $\ln(1 + x)$ in 1668. Both Newton (1665) and Gregory (1670) independently discovered the general binomial formula

$$(1 + x)^r = 1 + rx + \binom{r}{2} x^2 + \binom{r}{3} x^3 + \cdots$$

and made significant use of it, Newton in developing integration formulas and Gregory in series expansions. Taylor's theorem on the expansion of a function in a power series was first discovered by Gregory in 1671, although in a different form using differences. [See Stillwell (1989, Chapter 9).]

Power Series

Generalizing the geometric series $\sum ar^n$ by replacing r with a variable and a with a sequence $\{a_n\}$ is a natural step.

Definition 1.17 (Power Series) *A* power series centered at c *is a series having the form*

$$\sum_{n=0}^{\infty} a_n(x-c)^n = a_0 + a_1(x-c) + a_2(x-c)^2 + \cdots$$

where the a_n are constants.

By convention, $a_0(x-c)^0 = a_0$ for all values of x even though 0^0 is indeterminate. Suppose a power series converges for some x that is R units from c; then the comparison test tells us that the series converges for any x closer to c, i.e., for all x with $|x - c| < R$. *Prove it!* Similarly, if a power series diverges for some x that is R units from c, then it diverges for all x with $|x - c| > R$. This observation leads us to define the *radius of convergence* of a power series.

Definition 1.18 (Radius of Convergence) *For a power series $\sum a_n(x-c)^n$, exactly one of the three following statements must be true.*

1. *The series converges only at $x = c$. The radius of convergence is $R = 0$.*

2. *The series converges for all $|x - c| < R$ and diverges for all $|x - c| > R$ for some positive value R. The radius of convergence is R.*

3. *The series converges for all x. The radius of convergence is $R = \infty$.*

Given a positive radius of convergence R for the power series $\sum a_n(x - c)^n$, the *interval of convergence* is the interval from $c - R$ to $c + R$ that may or may not contain the endpoints—they must both be checked. Often, the radius of convergence can be found using the ratio or root tests. However, both tests fail at an endpoint of the interval of convergence, and other methods are required.

Mercator's expansion of $\ln(1 + x)$ comes from integrating the geometric series expansion for $1/(1 + x)$. Integrating

$$\frac{1}{1+x} = 1 - x + x^2 - x^3 \pm \cdots$$

yields

$$\ln(1+x) = x - \frac{x^2}{2} + \frac{x^3}{3} - \frac{x^4}{4} \pm \cdots$$

Newton investigated integrating series expansions for functions like $\sqrt{1 + y'^2}$ to find arc lengths without questioning convergence. Fortunately, power series convergence is preserved by differentiation and integration.

Theorem 1.40 *If $f(x) = \sum_{n=0}^{\infty} a_n(x - c)^n$ converges for $|x - c| < R \leq \infty$, then*

1. $f'(x) = \sum_{n=1}^{\infty} n\, a_n(x - c)^{n-1}$ *converges for $|x - c| < R$ and*

2. $\int_c^x f(t)\, dt = \sum_{n=0}^{\infty} a_n \dfrac{(x - c)^{n+1}}{n + 1}$ *converges for $|x - c| < R$.*

Let's explore techniques for finding power series expansions.

◼ **EXAMPLE 1.12**

Given that

$$\frac{1}{1 - x} = \sum_{n=0}^{\infty} x^n \qquad |x| < 1$$

find power series expansions for the following functions.

1. $f(x) = \dfrac{1}{1 + x^2}$

Replace x in the geometric power series with $-x^2$ to have

$$f(x) = \sum_{n=0}^{\infty} (-x^2)^n = \sum_{n=0}^{\infty} (-1)^n x^{2n} \qquad |x| < 1$$

2. $g(x) = x^{-1}$

Rewrite x^{-1} as $1/(1 + (x - 1))$. As above, replace x with $-(x - 1)$ in the geometric power series. Note the shift to center $c = 1$:

$$g(x) = \sum_{n=0}^{\infty} (-(x - 1))^n = \sum_{n=0}^{\infty} (-1)^n (x - 1)^n \qquad |x - 1| < 1$$

3. $h(x) = \dfrac{1}{(1 - x)^2}$

Since h is the derivative of $1/(1 - x)$, differentiate the geometric power series termwise:

$$h(x) = \sum_{n=0}^{\infty} \frac{d}{dx} x^n = \sum_{n=1}^{\infty} n x^{n-1} \qquad |x| < 1$$

4. $j(x) = \tan^{-1}(x)$

The arctangent is the integral of $1/(1 + x^2)$, the first series above:

$$j(x) = \sum_{n=0}^{\infty} \int (-1)^n x^{2n}\, dx = \sum_{n=0}^{\infty} (-1)^n \frac{x^{2n+1}}{2n + 1} \qquad |x| < 1$$

How does the constant of integration come into this computation? ◼

In 1772, Lagrange attempted to fix the "ghost of departed quantities" problem in the definition of derivatives and integrals by assuming that all functions could be represented as power series. [See Boyer (1959, p. 252) and Burton (2007, p. 525).] Unfortunately, not all functions have power series representations. The classic example of a function without a power series that has a nonzero radius of convergence is $f(x) = \exp(-1/x^2)$ for $x \neq 0$ and $f(0) = 0$. Taylor and Maclaurin studied the questions of when does a function have a power series representation and how to compute one.

Taylor Series

In 1715, Taylor published *Methodus Incramentorum Directa et Inversa* [see Struik (1986, p. 328)], which contained the result we now call Taylor's theorem. Maclaurin's 1742 text *Treatise on Fluxions* used the approach that we call "order of contact" to develop these power series. Since Maclaurin concentrated on series centered at $c = 0$, we call them Maclaurin series. The order of contact method matches derivatives of the function to derivatives of the power series in order to find the power series' coefficients. [See Burton (2007, p. 526).]

Theorem 1.41 (Taylor's Theorem) *Let f be a function that is continuous together with its first $n + 1$ derivatives on an interval containing c and x. Then*

$$f(x) = \sum_{k=0}^{n} \frac{f^{(k)}(c)}{k!}(x - c)^k + R_n(c, x)$$

where

$$R_n(c, x) = f^{(n+1)}(a) \frac{(x - c)^{n+1}}{(n + 1)!}$$

for some a between c and x.

We can rephrase the theorem in terms of power series.

Theorem 1.42 (Taylor Series) *If f has a power series expansion that is valid on $|x - c| < R$, that is, if*

$$f(x) = \sum_{k=0}^{\infty} a_k(x - c)^k \qquad |x - c| < R$$

then

$$a_k = \frac{f^{(k)}(c)}{k!}$$

■ **EXAMPLE 1.13   Important Maclaurin Series**

$$\frac{1}{1-x} = \sum_{k=0}^{\infty} x^k \qquad\qquad |x| < 1$$

$$\sin(x) = \sum_{k=0}^{\infty} (-1)^k \frac{x^{2k+1}}{(2k+1)!} \qquad\qquad x \in \mathbb{R}$$

$$\cos(x) = \sum_{k=0}^{\infty} (-1)^k \frac{x^{2k}}{(2k)!} \qquad\qquad x \in \mathbb{R}$$

$$\tan^{-1}(x) = \sum_{k=0}^{\infty} (-1)^k \frac{x^{2k+1}}{2k+1} \qquad\qquad |x| < 1$$

$$e^x = \sum_{k=0}^{\infty} \frac{x^k}{k!} \qquad\qquad x \in \mathbb{R}$$

$$\ln(1+x) = \sum_{k=1}^{\infty} (-1)^{k-1} \frac{x^k}{k} \qquad\qquad |x| < 1$$

■

To find the radius of convergence of a Taylor series, elementary calculus students usually use the ratio or root test. Typical examples of the computations are shown for e^x and $\ln(1+x)$.

■ **EXAMPLE 1.14**

1. Find the interval of convergence of the Maclaurin series for e^x.
   The general term of the series is $x^n/n!$, so the limit of the ratios is

$$r = \lim_{n\to\infty} \left| \frac{x^{n+1}/(n+1)!}{x^n/n!} \right| = \lim_{n\to\infty} \frac{|x|}{n+1} = 0$$

   Hence, the series converges for all x and the interval of convergence is $\mathbb{R}$.

2. Find the interval of convergence of the Maclaurin series for $\ln(1+x)$.
   The general term of the series is $(-1)^{n-1}x^n/n$, so the limit of the ratios is

$$r = \lim_{n\to\infty} \left| \frac{x^{n+1}/(n+1)}{x^n/n} \right| = \lim_{n\to\infty} \left| \frac{n}{n+1} \cdot x \right| = |x|$$

   Hence, the series converges for all $|x| < 1$. Since the function is undefined at $x = -1$, the series cannot converge there. By the alternating series test, the series does converge at $x = 1$. Thus the interval of convergence is $(-1, 1]$. ■

The ratio and root tests work for a large number of cases. However, there is more subtlety than is readily apparent. Consider the function $f(x) = \tan(\sin(x))$. It is quite difficult to find a simple expression for the general term of f's Maclaurin series. Naive reasoning suggests that since the sine is bounded by one and the tangent behaves quite well on $[-1, 1]$, there should be no problem for any x, no matter how large. Unfortunately, it's not that simple. Try as we might, we can never get convergence for $|x| > 1.9$. See Figure 1.7. What's happening here? The answer requires complex variables. When $z = \pi/2 - i \ln\left((\pi + (\pi^2 - 4)^{1/2})/2\right)$, we have $\sin(z) = \pi/2$, and hence $f(z) = \tan(\sin(z))$ is undefined. But that value of z is a complex number and we're only using real values! That doesn't matter. The radius of convergence is the distance to the nearest singularity, real or complex. And so, for $f(z) = \tan(\sin(z))$, we can never reach beyond $|z| \approx 1.87$ with a Maclaurin series, even on the real axis.

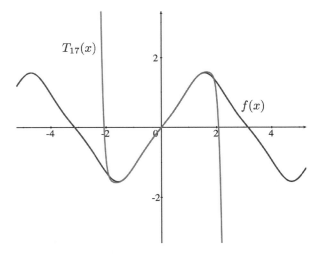

Figure 1.7 Graph of $f(x) = \tan(\sin(x))$ and $T_{17}(x)$

The delicateness of the radius of convergence question notwithstanding, Taylor and Maclaurin series provide powerful tools for analysis. Not only can we use the series very effectively for numerical approximation of functions that are difficult to compute, but the series also provide important theoretical tools. Taylor series are used in proofs throughout the theory of numerical analysis. Maclaurin series can be used as an integral part of the proof of the Weierstrass approximation theorem: *Let f be a continuous function on the closed interval $\mathcal{I}$. Given $\epsilon > 0$, there is a polynomial $p(x)$ such that $|f(x) - p(x)| < \epsilon$ for all $x \in \mathcal{I}$.*

Summary

We have recapitulated the standard topics of an elementary calculus course in this chapter. The order that concepts are presented here and in current courses is nearly opposite to their historical development. It took over two hundred years from the time that Newton and Leibniz made the crucial connection between derivative and integral to the limit-based definitions that are taught today. Nevertheless, it is remarkable that the origins of the limit concept can be traced to Eudoxus and even further back to Antiphon's bounds for π from the fifth century BC. What is equally surprising to students is that calculus is not fossilized and moribund but is still an active area of research.

EXERCISES

1.1 Let $f : X \to Y$ with $A, B \subseteq X$ and $C, D \subseteq Y$. Prove:

    a) $f(A \cap B) \subseteq f(A) \cap f(B)$
    b) $f(A \cup B) = f(A) \cup f(B)$
    c) $f\left(f^{-1}(C)\right) \subseteq C$
    d) $A \subseteq f^{-1}(f(A))$

1.2 Give examples of functions and sets for which

    a) $f\left(f^{-1}(C)\right) \neq C$
    b) $A \neq f^{-1}(f(A))$

1.3 Who was the first to write $f(x)$ to denote a function?

1.4 Describe the behavior of the two functions $f(x) = |ax + b|$ and $g(x) = |a|x + |b|$ for various values of a and b.

1.5 Compare $y = \sin(\arcsin(x))$ and $z = \arcsin(\sin(x))$.

1.6 Prove the *triangle inequality:* For $x, y \in \mathbb{R}$,

$$|x + y| \leq |x| + |y|$$

1.7 Prove the *inverse triangle inequality:* For $x, y \in \mathbb{R}$,

$$\big||x| - |y|\big| \leq |x| - |y|$$

1.8 Who first used the notation

$$\lim_{x \to a} f(x)$$

1.9 Determine the value of

$$\lim_{x \to 0} \frac{\sqrt{x^2 - a^2} - a}{x^2}$$

    a) for any real number $a > 0$,
    b) for any real number $a < 0$.

1.10 Find the value of

$$\lim_{x \to 0} (1 + nx)^{(1/x)}$$

for $n > 0$.

1.11 Using $\epsilon = \frac{1}{2}$, show that there is no $\delta > 0$ satisfying the definition of the limit for the unit step function at $a = 0$. Conclude that $\lim_{x \to 0} U(x)$ does not exist. (*Hint:* Suppose the limit is L. For any $\delta > 0$, choose $x_0 \in (0, \delta)$. Consider both $|U(x_0) - L|$ and $|U(-x_0) - L|$.)

1.12 The limit

$$\lim_{\theta \to 0} \frac{\sin(\theta)}{\theta} = 1$$

is important in calculus.

    a) Use the sandwich theorem to prove this limit.
    b) Use a geometric argument based on areas and the unit circle to establish this limit.

1.13 Calculate the limit

$$\lim_{\theta \to 0} \frac{1 - \cos(\theta)}{\theta}$$

1.14 Compute

$$\lim_{y \to 1} \frac{y^4 - 1}{y - 1}$$

1.15 Find

$$\lim_{x \to 3} \frac{\sqrt{x + 22} - 5}{x - 3}$$

a) by graphing,

b) by rationalizing the numerator.

1.16   Create a set of limit problems for an elementary calculus class illustrating the main techniques.

1.17   Create a set of limit problems for an elementary calculus class illustrating the main pitfalls.

1.18   Compare and contrast the approaches used in standard calculus textbooks such Granville et al. (1911), Thomas (1968), Finney et al. (1999), Ostebee & Zorn (2002), and Stewart (2009) to the limit concept.

1.19   For each of the three conditions listed for the definition of continuity, give examples that fail to satisfy that condition.

1.20   Show that a polynomial is continuous everywhere.

1.21   Give an example for each kind of the four types of discontinuity listed in Table 1.1.

1.22   Is the function

$$f(x) = \begin{cases} e^x - x - 1 & x > 0 \\ 0 & \text{otherwise} \end{cases}$$

continuous at $x = 0$?

1.23   Discuss the continuity of

$$g(x) = x - \lfloor x \rfloor$$

1.24   Choose a value of a so that

$$h(x) = \begin{cases} x^2 - ax - 1 & x \leq 1 \\ x^3 - x^2 + a & x > 1 \end{cases}$$

is continuous everywhere.

1.25   Determine the points of continuity of the function

$$k(z) = \begin{cases} e^z - 1 & z > 0 \\ 0 & \text{otherwise} \end{cases}$$

1.26   Let $r(x) = p(x)/q(x)$ be a rational function. If the degree of p is n and that of q is m, what is the largest number of discontinuities that r can have? The smallest?

1.27   If f is continuous at a and g is not, then $f + g$ is not continuous at a. Can $f + g$ be continuous at a when neither f nor g is continuous at a?

1.28   Prove that the composition of continuous functions is continuous (Theorem 1.8) by using Theorem 1.7.

1.29   Give examples showing that each hypothesis, continuous and closed interval, is necessary in the intermediate value theorem.

1.30   Compare and contrast the approaches used in standard calculus textbooks like Granville et al. (1911), Thomas (1968), Anton et al. (2009), Finney et al. (1999), Ostebee & Zorn (2002), and Stewart (2009) to continuity.

1.31   Write careful definitions for left- and right-hand derivatives at a point.

1.32   Let $f(x) = \sqrt{x + 22}$.
   a) Compute $f'(3)$.
   b) Compute $f''(3)$.
   c) Compute $f'(x)$.
   d) Compute $f''(x)$.

1.33   Explain why any tangent line to a parabola intersects the curve exactly once.

1.34   Find the derivative of

$$g(x) = \ln\left(x + \ln\left(x + \ln(x)\right)\right).$$

1.35   Compute the derivative of

$$h(t) = \cfrac{1}{1 + \cfrac{1}{1 + t}}$$

a) without first simplifying the complex fraction,

b) after simplifying the complex fraction.

1.36 Let $k(x) = x^3 - 3x + 2$.
a) Find the equation of the line L tangent to k at $x = 1/2$.
b) Find all intersection points of L and k.
c) Graph L and k together.

1.37 Let $k(x) = x^3 - 3x + 2$.
a) Find the equation of the line N normal to k at $x = -1$.
b) Find all intersection points of N and k.
c) Graph N and k together.

1.38 Write the derivative rules
a) for the six trigonometric functions,
b) for the six inverse trigonometric functions.

1.39 Using the convention $f^{(0)} = f$, prove

$$\cos^{(n)}(x) = \cos^{(n \bmod 4)}(x)$$

for all $n \in \mathbb{N}$.

1.40 Consider logarithmic derivatives:
a) Execute the following four Maple statements several times.
```
r := [seq(rand(-9..9)(), k=1..4)];
p := expand(
        product(x-r[k], k=1..4));
dp := diff(ln(p), x):
dp = convert(dp, parfrac, x)
```
What do you observe?
b) Let $p(x)$ be a polynomial. Compute

$$\frac{d}{dx} \ln(p(x))$$

1.41 Differentiate $y = x^x$.

1.42 What is the minimum value of the function

$$f(x) = x + \frac{1}{x}$$

1.43 The distance by bus from New York to Boston is 215 miles. A bus driver gets paid \$19.50 per hour. The cost of running the bus at a steady speed of r miles per hour is $0.80 + 0.005r$ dollars per mile. The minimum and maximum legal speeds on the roads are 40 and 55 miles per hour. What steady speed minimizes the total cost of a nonstop trip?

1.44 A farmer wishes to fence a rectangular field and to divide the field in half with another fence. The outside fence costs \$4 per foot, and the fence in the middle costs \$3 per foot. If the farmer has budgeted \$1000 for fencing, what dimensions maximize the total area?

1.45 A cylindrical grain bin of radius 10 feet and height 30 feet is being filled with corn at the rate of 30 cubic feet per minute. How fast is the depth of the corn increasing?
a) Assume the corn is spread in a uniform layer.
b) Assume the corn falls in a conical pile constrained by the bin walls.

1.46 Compare and contrast the approaches used in standard calculus textbooks like Granville et al. (1911), Thomas (1968), Finney et al. (1999), Ostebee & Zorn (2002), and Stewart (2009) to derivatives.

1.47 Use a computer algebra system to generate a random, fifth-degree polynomial $p(x)$. Define *Newton's method func-*

tion $N(x)$ by

$$N(x) = x - \frac{p(x)}{p'(x)}$$

a) Find the smallest positive root r by iterating N, starting with an initial guess taken from a graph.

b) Define a new iterating function

$$M(x) = x - \frac{p'(x)}{p''(x)}$$

What is the result of iterating M with the starting value r found in part a?

1.48 Use a geometric argument to show that

$$\int_{-r}^{r} \sqrt{r^2 - x^2}\, dx = \frac{1}{2}\pi r^2$$

for $r > 0$.

1.49 Compute

a) $\displaystyle\int \frac{\sqrt{x} + 3x^3}{x}\, dx$

b) $\displaystyle\int \frac{\sin(\sqrt{z})}{\sqrt{z}}\, dz$

c) $\displaystyle\int \frac{t}{\sqrt{t+1}}\, dt$

1.50 Compute

a) $\displaystyle\int_{-1}^{1} 1 - x^2\, dx$

b) $\displaystyle\int_{0}^{\pi} \frac{\sin(x)}{x}\, dx$

c) $\displaystyle\int_{0}^{1} 2\pi \left(\sqrt{x} - x^2\right)^2 dx$

1.51 Find an equation for the curve passing through the point $(2, 1)$ and having slope $y' = 2x^3 - x - 3$ at each point (x, y).

1.52 Find the geometric area between the sine and cosine curves from 0 to 2π.

1.53 Use a computer algebra system to integrate

$$\int \sin\left(\frac{\pi}{2} t^2\right) dt$$

Investigate the result.

1.54 Let f be integrable on $[a, b]$ and be bounded above by M and below by m. Prove

$$m(b - a) \le \int_{a}^{b} f(x)\, dx \le M(b - a)$$

1.55 Consider the function

$$L(x) = \frac{x^{n+1} - 1}{n + 1}$$

a) Show that $L'(x) = x^n$ and conclude that L is an antiderivative for x^n.

b) Calculate

$$\lim_{n \to -1} L(x)$$

How does this result relate to part a?

1.56 Use a graphing utility to construct an accurate "three-rectangle diagram" showing upper and lower bounds and the average value for the integral $\int_0^3 e^{-x^2}\, dx$.

1.57 Use a computer algebra system or a calculator to make the counterpart of Table 1.2 for the integral $\int_0^\pi \sin(x)/x\, dx$.

1.58 Compare and contrast the approaches used in standard calculus textbooks like Granville et al. (1911), Thomas (1968), Anton et al. (2009), Finney et al. (1999), Ostebee & Zorn (2002), and Stewart (2009) to the concept of integration.

1.59 Prove Theorem 1.32, the algebra of sequence limits.

1.60   Determine whether the given sequence converges.

a) $a_n = (-1)^n \cdot \dfrac{n-1}{n}$

b) $b_n = \dfrac{n^2}{2^n}$

c) $c_n = \dfrac{n!}{2 \cdot 4 \cdot 6 \ldots (2n)}$

1.61   Determine whether the given recursively defined sequence converges.

a) $p_n = \sqrt{2 + p_{n-1}}$ for $n > 1$ and $p_1 = \sqrt{2}$

b) $q_n = \dfrac{q_{n-1} + q_{n-2}}{2}$ for $n > 2$ with $q_1 = 0$ and $q_2 = 1$

c) $f_n = f_{n-1} + f_{n-2}$ for $n > 2$ with $f_0 = 1$ and $f_1 = 1$

1.62   Suppose that a_n converges to 0. Does the sequence

$$b_n = \sin\left(\frac{n\pi}{2}\right) \cdot a_n$$

converge or diverge?

1.63   Suppose that c_n converges to 1. Does the sequence

$$d_n = \sin\left(\frac{n\pi}{2}\right) \cdot c_n$$

converge or diverge?

1.64   Investigate the Maclaurin series for

a) $\tan(x)$,

b) $\sec(x)$.

1.65   What value of the center a gives the best Taylor fourth-degree polynomial approximation to $y = e^x$ on the interval $[-1, 1]$?

1.66   Explain why the following functions do not have Maclaurin series.

a) $f(x) = |x|$

b) $g(x) = \ln(x)$

c) $h(x) = \cot(x)$

d) $k(x) = \sqrt{x}$

1.67   Find the Taylor expansion for

$$f(x) = \frac{1}{\sqrt{1 - x^2}}$$

centered at $a = 0$.

1.68   Let T_n be the nth-degree Maclaurin polynomial of arcsin .

a) Find T_n for $n = 10$.

b) Plot T_n and arcsin together for $n = 3, 6, 10$.

1.69   Let $f(x) = \tan(\cos(x))$.

a) Produce graphs of several Taylor polynomials of f for different choices of n. What do you observe?

b) Using a 3D graphing utility, plot $z(x, y) = |f(x + iy)|$ in the window $[-2, 2] \times [-2, 2]$. How does this relate to part a?

1.70   Let i be the imaginary unit $\sqrt{-1}$.

a) Calculate all the integer powers of i.

b) Substitute ix in the Maclaurin expansion of e^x.

c) Separate the expansion of e^{ix} into terms with i and terms without.

d) Compare the results above with the Maclaurin expansions of sine and cosine.

e) Write Euler's identity

$$e^{ix} = \ldots .$$

1.71   Who said,

I seem to have been like a child playing on the sea shore, finding now and then a prettier shell than ordinary, whilst the great ocean of truth lay undiscovered before me.

1.72   Who said,

I hold that the mark of a genuine idea is that its possibility can be proved, either a priori by conceiving its cause or reason, or a posteriori when experience teaches us that it is in fact in nature.

Now that you have finished looking back at calculus, it's time to begin analysis. There is one last assignment, however—reflect on Morris Kline's comment:

Contrary to common belief, the calculus is not the height of the so-called "higher mathematics." It is, in fact, only the beginning.

INTERLUDE: FERMAT, DESCARTES, AND THE TANGENT PROBLEM

If we were to teach calculus following the order that topics were originally developed, we would start with integration, then study differentiation, and finally consider limits—the exact opposite of modern classes.

Eudoxus of Cnidus (408–355 B.C.) based his results concerning areas and volumes on his *method of exhaustion*. The term "exhaustion" refers to the difference in area or volume of a given object to the approximating regular figures being "exhausted" or "used up." Eudoxus would recognize the illustrations of approximating sums for a definite integral shown in modern calculus texts. The method of exhaustion was the main tool used to prove quadrature formulas (our definite integrals) up to the time of Fermat. Cavalieri had developed formulas for the integral of x^n for x from 0 to a using "indivisibles" in his 1635 text *Geometria indivisibilibus*. These indivisibles became Newton's *moments* and Leibniz's *differentials*. Cavalieri's book was extremely influential and widely read by mathematicians of the 1600s [see O'Connor & Robertson (2008)].

The tangent problem—construct a line tangent to a curve at a specified point—is the generalization arising from Euclid's definition of the tangent to a circle. Archimedes gave methods to construct tangents to spirals and other curves. Descartes, in his *Géométrie*, considered "algebraic curves," or curves with simple algebraic formulas, and rejected "mechanical curves." While Descartes used algebra heavily, his approach was really based on geometry. Fermat used algebra with infinitesimals, possibly based on Cavalieri's indivisibles, and approached the tangent line problem in a fashion similar to what we do in classes today.

Let's consider the problem of finding the tangent line to the curve $y^2 = 3x$ at the point $(3, 3)$ using both Descartes' and Fermat's methods.

Descartes' Tangent Circle

Descartes' technique was to find a circle centered on the x-axis that was tangent to the curve at the point in question and then recognize that the normal to the radius at that point forms the tangent to the original curve. See Figure I.1.

Begin by looking at the equation of a circle centered on the x-axis at $(h, 0)$ and passing through the point $(3, 3)$. The circle's radius is $r = \sqrt{(3 - h)^2 + 3^2}$. Hence,

$$(x - h)^2 + y^2 = h^2 - 6h + 18$$

Substitute $y^2 = 3x$ to have the curve and circle intersect. Apply a little algebra to obtain

$$x^2 - 2xh + 3x + 6h - 18 = 0$$

A crucial observation is that the circle is tangent to the curve when the equation above has a single root, i.e., a single intersection in the neighborhood of 3. Since the roots are 3 and $2h - 6$, we take $h = 9/2$. Descartes' computation reduces the

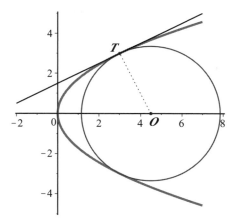

Figure I.1   Descartes Tangent Circle

problem to the simple task of finding a line passing through $(3, 3)$ and perpendicular to the segment joining $(9/2, 0)$ to $(3, 3)$. The desired tangent is easily calculated to be $y = 1/2 \cdot (x - 3) + 3$.

Note that Descartes does not use limits—even implicitly—in his calculations.

Fermat's Similar Triangles

Fermat's technique is based on similar triangles. When the triangles coincide, the hypotenuse lies on the tangent line. See Figure I.2.

Fermat starts by drawing the line tangent to the curve at $(3, 3)$. He then creates triangle $\triangle OAT$ by dropping a line to the x-axis. Add the length E to point A to create a new triangle $\triangle OBP$. Set the length of the segment $\overline{OA}$ to be s; then the length of $\overline{OB}$ is $s + E$. Since the two triangles are similar, the ratios of the legs are equal. At this point, Fermat substitutes $f(3 + E) = \sqrt{3(3 + E)}$ for the length of $\overline{BP}$, claiming that the error disappears when $E = 0$. Thus

$$\frac{s}{s + E} = \frac{\sqrt{3 \cdot 3}}{\sqrt{3(3 + E)}} = \frac{3}{\sqrt{3(3 + E)}}$$

Solve this expression for s. *Do the algebra!*

$$s = \frac{3}{\left(\sqrt{3(3 + E)} - 3\right)/E} = \sqrt{3(3 + E)} + 3$$

With $E = 0$, we have $s = 6$.

The critical observation is that the slope of the tangent line is equal to the length of $\overline{AT}$ divided by s. Hence the slope is $3/6 = 1/2$. The equation of the tangent line is given by $y = 1/2 \cdot (x - 3) + 3$.

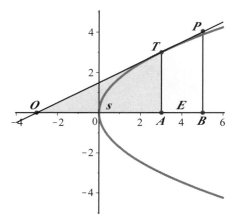

Figure I.2   Fermat's Similar Triangles

Conclusion

Descartes' technique did not involve using infinitesimals and letting values not be zero for simplification, then be zero for completion. Because he relied on standard algebraic manipulation, Descartes' approach could be rigorously defended. However, his method was limited to simple algebraic curves by the nature of the computations.

Fermat's technique implicitly uses a limiting process as E goes to zero. Even though there was no contemporary logical justification, Fermat's approach handled a wide variety of curves and produced results that were correct.

Unfortunately, Descartes and Fermat became entangled in a controversy over priority. Finally, Descartes admitted that Fermat's method was the more general, but the enmity between them was never overcome.

CHAPTER 2

INTRODUCTION TO REAL ANALYSIS

For Mathematical Proofs, like Diamonds, are hard as well as clear, and will be touch'd with nothing but ſtrict Reaſoning.

—John Locke

Introduction

We begin our study of real analysis by reviewing the basic properties of the real numbers and the real line. The set of real numbers is so familiar to us that many significant properties are taken for granted, but we need to look at them carefully. Our plan in this chapter is to lay the foundations of real analysis, thus making calculus rigorous. We'll consider essentially the same topics as Chapter 1, limits and continuity, differentiation, integration, and sequences and series, but now from a perspective based on logical development and proof. Historically, we are following the growth of real analysis from Cauchy to Riemann. According to Boyer (1959), this was the period of "Rigorous Formulation" in calculus development.

Opening quotation is from *Mr Locke's Reply to the Right Reverend Lord Bishop of Worcester's Answer to his Second Letter.* c.1669.

Introduction to Real Analysis. By William C. Bauldry
Copyright © 2009 John Wiley & Sons, Inc.

Going forward, we must keep in mind the advice of Arnold E. Ross, founder of the Ross Mathematics Program, who paraphrased Gauss in choosing the Ross program's motto "think deeply of simple things."

2.1   BASIC TOPOLOGY OF THE REAL NUMBERS

Since the real numbers underlie all of our work, we start by examining their basic properties. Many of these properties are quite familiar and seem self-evident at first glance, but we must still prove their validity.

Fields

The real numbers $\mathbb{R}$ with the operations $+$ and $\times$ comprise a *field*. In general, a field is a set with two operations that satisfy the eleven axioms listed in Table 2.1.

1a. Addition is closed	1m. Multiplication is closed
2a. Addition is associative	2m. Multiplication is associative
3a. Addition is commutative	3m. Multiplication is commutative
4a. There is an additive identity	4m. There is a multiplicative identity
5a. Every number has an additive inverse	5m. Every nonzero number has a multiplicative inverse
6. Multiplication distributes over addition	

Table 2.1   The Field Axioms

Standard examples of fields include the rational numbers $\mathbb{Q}$, the real numbers $\mathbb{R}$, and the complex numbers $\mathbb{C}$, all with the usual $+$ and $\times$. *Verify this!* The integers $\mathbb{Z}$ fails only one of the eleven axioms and so $\mathbb{Z}$ is not a field. *Which one is it?*

The uniqueness of the identity elements 0 and 1 is an immediate consequence of the axioms. To wit: If 0 and $0'$ are both additive identities, then $0 = 0 + 0' = 0'$, hence $0 = 0'$, etc. The uniqueness of additive and multiplicative inverses also follows directly. Many important arithmetic results are easy applications of the field axioms. For example, the *cancellation laws* are directly implied by the existence of inverses and associativity.

Theorem 2.1 (Cancellation Laws) *Let x, y, and z be real numbers.*

1. *If $x + y = x + z$, then $y = z$.*

2. *If $xy = xz$ and $x \neq 0$, then $y = z$.*

This proof[1] is left to the exercises.

---

[1] Barnard's "Why are proofs difficult?" Barnard (2000) provides a very good discussion of "proof" from a pedagogical viewpoint.

Ordered Sets

A set is *ordered* when it has a transitive relation $<$ such that any two elements are comparable. That is, the two statements

1. For each x and y, exactly one of $x < y$, $x = y$, or $y < x$ is true.

2. For any x, y, and z, if $x < y$ and $y < z$, then $x < z$.

must be true. For the real numbers, the first statement is called the *trichotomy law*. The rational numbers $\mathbb{Q}$ and the real numbers $\mathbb{R}$ with our usual $<$ relation are ordered sets. There is not a simple order relation for the complex numbers $\mathbb{C}$.

Once we have ordered a set, we can consider upper and lower bounds.

Definition 2.1 (Least Upper Bound Property) *Let U be an ordered set. Then U has the* least upper bound property *if, whenever a nonempty subset $E \subseteq U$ is bounded above, then E has a least upper bound in U. We call the least upper bound of E the* supremum of E *and write* $\sup E$.

The real numbers $\mathbb{R}$ ordered by $<$ possess the least upper bound property. The rational numbers $\mathbb{Q}$ with $<$ do not have the the least upper bound property. *Give an example to show this!* The least upper bound property is also called the *completeness property*. The reason for the term "completeness" is that this property eliminates the possibility of gaps in the real line. Suppose there was a gap at some value $\hat{x}$. Then the set $A = \{x \in \mathbb{R} \mid x < \hat{x}\}$ is bounded above by $\hat{x} + 1$. However $\sup A = \hat{x}$ and $\hat{x} \notin \mathbb{R}$ contradicts the least upper bound property. Therefore there can be no gaps in the real line.

There is a natural relation between least upper bounds and greatest lower bounds.

Theorem 2.2 *Let U be an ordered set possessing the least upper bound property. Suppose a nonempty subset $B \subseteq U$ is bounded below. Then the greatest lower bound of B exists and is in U. We call the greatest lower bound of E the* infimum of E *and write* $\inf E$.

Proof: Set L to be the set of all lower bounds of B. Then $L \neq \emptyset$ because B is bounded below. Since every $b \in B$ is an upper bound for L, we have that L is bounded above. By the least upper bound property, the least upper bound β of L must be in U.

If $\alpha < \beta$, then α is not an upper bound for L; thus $\alpha \notin B$. Therefore $\beta \leq b$ for every $b \in B$, which makes β a lower bound for B. Hence, $\beta \in L$.

If $\alpha > \beta$, then $\alpha \notin L$, since β is an upper bound for L. Thus α is not a lower bound for B.

These two statements taken together demonstrate that β is the greatest lower bound for B; i.e., $\beta = \inf B$ and $\beta \in U$. ∎

If an ordered field has the least upper bound property, it is called a *complete ordered field*. Hence, the real numbers $\mathbb{R}$ with $+$, $\times$, and the relation $<$ form a complete ordered field.

The *Archimedean order property* is a simple, but very useful, result that follows easily from the least upper bound property.

Theorem 2.3 (Archimedean Order Property) *If ϵ and x are positive real numbers, then there is an $n \in \mathbb{N}$ such that $n\epsilon > x$.*

Proof: Suppose no such n exists. Then the set $A = \{\epsilon, 2\epsilon, 3\epsilon, \dots\}$ is bounded above by x. By the least upper bound property, $\beta = \sup A$ exists and $\beta \in \mathbb{R}$. Then every multiple of ϵ must be less than β. In particular, for any n, we have $(n+1)\epsilon \leq \beta$ or $n\epsilon \leq \beta - \epsilon$. However, this inequality indicates that $\beta - \epsilon$ is an upper bound for A which is smaller than β, the least upper bound—a contradiction. Therefore an n such that $n\epsilon > x$ must exist.   ∎

There are a number of equivalent forms of the Archimedean order property.

Corollary 2.4 *The following are equivalent.*

1. *The Archimedean order property.*

2. *For any $x \in \mathbb{R}$, there is an $n \in \mathbb{N}$ such that $n > x$.*

3. *For any positive real number x, there is an $n \in \mathbb{N}$ such that $1/n < x$.*

4. *For any positive real number x, there is an $n \in \mathbb{N}$ such that $n \leq x < n+1$.*

The proof is left to the reader.

Another immediate consequence of the Archimedean order property is that every open interval contains both rational and irrational numbers. The proof will be left to the exercises.

Theorem 2.5 *Every open interval (a, b) contains both a rational and an irrational number.*

The above theorem indicates that the set of rationals $\mathbb{Q}$ is *dense in* $\mathbb{R}$. That is, given any interval $(a, b) \subseteq \mathbb{R}$ where $a < b$, the set $(a, b) \cap \mathbb{Q} \neq \emptyset$. Thus, the rationals are spread uniformly throughout the real numbers. The same is true of the irrationals.

Curiously, even though between any two irrational numbers there must be a rational, we'll see later that there are so many more irrationals that the rationals disappear into the background.

Metric Spaces and Open Sets

We have worked with the field and order properties of $\mathbb{R}$; now it's time to focus on a distance function and the forms of subsets to round out our picture of the structure of the real numbers.

Definition 2.2 (Metric Space) *A metric space is a set X with a distance function $d : X \to \mathbb{R}$ that satisfies*

1. $d(x, y) = 0$ *if and only if $x = y$;*

2. $d(x, y) = d(y, x)$;

3. $d(x, z) \leq d(x, y) + d(y, z)$ *(triangle inequality).*

The function d is called a metric.

From these three properties, we have that $0 = d(x, x) \le d(x, y) + d(y, x) = 2d(x, y)$. It follows that $d(x, y) \ge 0$. The most familiar examples of metric spaces are the *Euclidean n-spaces* $\mathbb{R}^n$ with the Euclidean distance $d(\vec{x}, \vec{y}) = \sqrt{\sum(x_i - y_i)^2}$. For $n = 1$, this metric reduces to $d(x, y) = |x - y|$. *Show that $|x - y|$ satisfies the properties of a distance function.*

We collect several definitions into a group that will allow us to define the structure of subsets of $\mathbb{R}$.

Definition 2.3 *Let X be a metric space with metric d.*

1. *A* neighborhood *of a point x is the set $N_r(x) = \{y \mid d(x, y) < r\}$ for some $r > 0$. A* deleted neighborhood *is $N_r'(x) = N_r(x) - \{x\}$.*

2. *A point x is an* interior point *of a set E if there is a neighborhood $N_r(x) \subseteq E$ for some $r > 0$.*

3. *The* interior *of a set E is the set of all interior points of E.*

4. *E is* open *if every point of E is an interior point.*

5. *A point x is an* accumulation point *or a* limit point *of E if every neighborhood $N_r(x)$ contains a point of E different from x; i.e., $N_r'(x) \cap E \ne \emptyset$.*

6. *E is* closed *if every point of E is an accumulation point of E.*

7. *The* closure *of a set E is the union of E and the set of all accumulation points of E.*

8. *A point $x \in E$ is an* isolated point *if x is not an accumulation point of E.*

In $\mathbb{R}$, a neighborhood of a point is an open interval centered about the point. Since an open interval contains a smaller open interval about any point in it, an open interval is an open set. If we add both endpoints to the interval, we have a closed set. Hence the names open and closed interval.

Several results follow directly from the definitions. The proofs are left to the reader.

Theorem 2.6 *Let X be a metric space with metric d.*

1. *Every neighborhood is open.*

2. *Every neighborhood of a limit point of E contains infinitely many points of E.*

3. *A finite set has only isolated points.*

4. *A set is open if and only if its complement is closed.*

5. *The interior of a set is an open set.*

6. *The closure of a set is a closed set.*

How do intersections and unions of open or closed sets behave? Very nicely in certain circumstances. Arbitrary unions and finite intersections of open sets stay open.

Arbitrary intersections and finite unions of closed sets stay closed. To see that an intersection of an infinite number of open sets can be closed consider the collection of open intervals $O_n = (-1 - 1/n, 1 + 1/n)$. Then $\bigcap_n O_n = [-1, 1]$, a closed interval. *Create an example to show that the union of an infinite collection of closed sets may be open.*

Theorem 2.7

1. *For any collection of open sets $\{O_\alpha\}$, the union $\bigcup_\alpha O_\alpha$ is open.*

2. *For any collection of closed sets $\{F_\alpha\}$, the intersection $\bigcap_\alpha F_\alpha$ is closed.*

3. *For any finite collection of open sets $\{O_n\}$, the intersection $\bigcap_n O_n$ is open.*

4. *For any finite collection of closed sets $\{F_n\}$, the intersection $\bigcup_n F_n$ is closed.*

Proof:

1. Set $O = \bigcup O_\alpha$. If $x \in O$, then there is an α such that $x \in O_\alpha$. Since O_α is open, there is a neighborhood $N_r(x) \subseteq O_\alpha$. But since O is a union, then $N_r(x) \subseteq O_\alpha \subseteq O$ and O is open.

2. Set O_α to be the complement of F_α and apply 1 using $\left(\bigcap_\alpha F_\alpha\right)^c = \bigcup_\alpha (F_\alpha^c)$ where F^c denotes the complement.

3. Let $O = \bigcap_{n=1}^{N} O_n$ and let $x \in O$. Then x is in each O_n. So there is a neighborhood $N_{r_n}(x) \subseteq O_n$ for each $n = 1, \ldots, N$. Choose $r = \min\{r_1, r_2, \ldots, r_N\}$, then $N_r(x)$ is contained in each O_n, and so in the intersection. Hence O is open.

4. Set F_n to be the complement of O_n and apply 3 using $\left(\bigcup_\alpha O_\alpha\right)^c = \bigcap_\alpha (O_\alpha^c)$.

∎

What is the structure of an open set in $\mathbb{R}$? By the results above, we see that an open set in $\mathbb{R}$ can be written as a disjoint union of open intervals. (See Exercise 2.8.)

As we noted above, the real numbers $\mathbb{R}$ with $d(x, y) = |x - y|$ form a metric space. Metric spaces also have a concept called completeness. A metric space is *complete* if and only if every Cauchy sequence converges. (See Definition 2.16.) This meaning of completeness also prevents gaps in the real line. If there were a "missing" point, we could easily construct a sequence converging to that point contradicting completeness.

We have assumed the existence of the set of real numbers and the basic arithmetic properties. For those who wish to work through a rigorous construction of the real numbers using *Dedekind cuts*, refer to Rudin (1976). To see a development of the real numbers using *Cauchy sequences* along with historical perspectives, read Hairer & Wanner (1996).

2.2   LIMITS AND CONTINUITY

Limits and continuity are two fundamental concepts in real analysis. While intuitively obvious, both concepts have subtleties that we must be careful to address. We now set aside the heuristics of elementary calculus—does the graph appear to connect, to have no holes—and carefully base our development on logic.

The Limit of a Function at a Point

In elementary calculus, we arrived at Weierstrass's "ϵ-δ" definition of the limit to make the idea of "closeness" precise. We required a function f to be defined on a deleted neighborhood of a point a before asking if f had a limit as x approached a. This requirement is too restrictive. What is the essential feature we need? The function f must have sufficiently many domain points near a to establish its behavior. If a is an accumulation point of the domain, then there are infinitely many other domain points nearby. We'll use this criterion in our definition of limit.

Definition 2.4 *Let a be an accumulation point of the domain of the function f. Then $\lim_{x \to a} f(x) = L$ if and only if given any $\epsilon > 0$ there is a $\delta > 0$ so that, whenever $x \in \mathrm{dom}(f)$ and $0 < |x - a| < \delta$, then $|f(x) - L| < \epsilon$.*

Before proceeding to examples, we show that a limit can take at most one value.

Theorem 2.8 *If the limit of $f(x)$ as x approaches a exists, then it is unique.*

Proof: Suppose that the limit exists taking on two values L and M, and let $\epsilon > 0$ be given. Find $\delta_L > 0$ such that if $x \in \mathrm{dom}(f)$ and $0 < |x - a| < \delta_L$ then $|f(x) - L| < \epsilon/2$. Also find $\delta_M > 0$ such that if $x \in \mathrm{dom}(f)$ and $0 < |x - a| < \delta_M$ then $|f(x) - M| < \epsilon/2$. Now let $\delta = \min\{\delta_L, \delta_M\}$. Then

$$|L - M| \leq |L - f(x)| + |f(x) - M| < \frac{\epsilon}{2} + \frac{\epsilon}{2} = \epsilon$$

for any $x \in \mathrm{dom}(f)$ with $0 < |x - a| < \delta$. Since ϵ was arbitrary, $L = M$. ∎

Let's look at a relatively simple example.

■ **EXAMPLE 2.1**

Verify the limit

$$\lim_{x \to 1} x^3 + x + 1 = 3$$

We first check to make sure that $a = 1$ is an accumulation point of the domain of $x^3 + x + 1$. Since the domain is $\mathbb{R}$, our point must be in it.

Let $\epsilon > 0$ be given. Analyzing $|f(x) - L|$ will help us to choose δ.

$$
\begin{aligned}
|(x^3 + x + 1) - 3| &= |x^3 + x - 2| \\
&= |x - 1| \cdot |x^2 + x + 2| \\
&\leq |x - 1| \cdot (|x|^2 + |x| + 2)
\end{aligned}
$$

The $|x - 1|$ term is controlled by δ. We must handle the other term separately. Since $|x - 1| < \delta$ is the same as $1 - \delta < x < 1 + \delta$, choosing $\delta \leq 1$ yields $0 < x < 2$. Thus $|x|^2 + |x| + 2 \leq 8$ for $x \in (0, 2)$, so that

$$|(x^3 + x + 1) - 3| \leq |x - 1| \cdot 8$$

This inequality tells us to choose $\delta < \epsilon/8$. Put these computations together to have that if $0 < |x - 1| < \delta$ then

$$|(x^3 + x + 1) - 3| \leq |x - 1| \cdot 8 < 8\delta < \epsilon$$

whenever δ is chosen so that $\delta < \min\{1, \epsilon/8\}$. ∎

Now, an example of a simple limit that fails to exist.

■ **EXAMPLE 2.2**

Show that the limit of $f(x) = 1/x$ does not exist at $a = 0$.
  Let $\epsilon = 1/2$. We will show that no $\delta > 0$ can give $|1/x - L| < 1/2$ for any L. Let $\delta > 0$. Choose $n \in \mathbb{N}$ such that $0 < 1/n < \delta$.
  Then, if the limit exists,

$$\left| f\left(\frac{1}{n}\right) - f\left(\frac{-1}{n}\right) \right| \leq \left| f\left(\frac{1}{n}\right) - L \right| + \left| L - f\left(\frac{-1}{n}\right) \right| < \frac{1}{2} + \frac{1}{2} = 1$$

However, $|f(1/n) - f(-1/n)| = 2n > 1$. This is a contradiction; therefore the limit cannot exist. ∎

The examples above are typical of those handled in a basic calculus course. Let's investigate a little more subtle example that was given in Borzellino's (2001) paper "Whose Limit Is It Anyway?"

■ **EXAMPLE 2.3** **Borzellino's Limit Example**

Let $f(x) = \sqrt{x \sin(1/x)}$ and determine the limit of $f(x)$ as x goes to 0.
  The domain of f has 0 as an accumulation point, so a limit as x approaches 0 is a possibility. The graph of f appears in Figure 2.1.
  Let $\epsilon > 0$ be given. Then choose $\delta \leq \epsilon^2$. Now let x be a point in the domain of f such that $|x| < \delta$. Since the sine is bounded by 1, we see that

$$|f(x) - 0| = \left| \sqrt{x \sin\left(\frac{1}{x}\right)} \right| \leq \sqrt{x} < \sqrt{\delta} \leq \epsilon$$

Thus $\lim_{x \to 0} f(x) = 0$. ∎

Since $x \sin(1/x)$ takes on negative values arbitrarily close to zero, those values are not in the domain of f. No neighborhood of $a = 0$ is a subset of the domain of f. The limit definition of Chapter 1 isn't applicable. The function of Example 2.3 cannot be considered in an elementary calculus course.

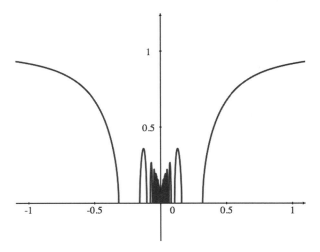

Figure 2.1 Graph of the Function $f(x) = \sqrt{x}\sin(1/x)$

■ EXAMPLE 2.4

Determine $\lim_{x \to 0} \sqrt{|x^2 - 1| - 1}$.

Figure 2.2 shows a graph of our function.

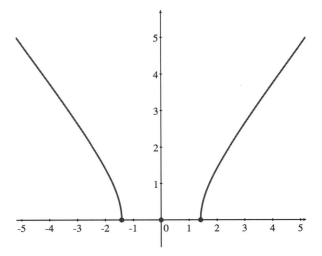

Figure 2.2 Graph of the Function $g(x) = \sqrt{|x^2 - 1| - 1}$

Note that 0 is in the domain of g and $g(0) = 0$. However, 0 is an isolated point of the domain, not an accumulation point. The definition of limit can't be applied. This limit problem is called "ill-posed" since the point we are attempting to approach is not an accumulation point of the domain.   ∎

Let's see one more example before moving on to the algebra of limits. A function can de defined everywhere, but have rather exotic behavior.

■ **EXAMPLE 2.5   Thomae's Function**

Define $T : [0, 1] \to \mathbb{R}$ by

$$T(x) = \begin{cases} \dfrac{1}{q} & x = \dfrac{p}{q} \in \mathbb{Q} \text{ in lowest terms} \\ 0 & x \notin \mathbb{Q} \end{cases}$$

For any $a \in [0, 1]$, we have $\lim_{x \to a} T(x) = 0$.
    Figure 2.3 shows a graph of Thomae's function.

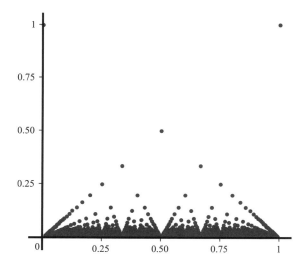

Figure 2.3   Graph of Thomae's Function on $[0, 1]$

The domain is a closed interval; therefore every point in it is an accumulation point; the limit is well-posed.
    Let $\epsilon > 0$ be given. For any irrational x, the function value is 0, whence $|T(x) - 0| = 0 < \epsilon$. So we need only consider rational values near a. There are only a finite number of rationals p/q for which $1/q > \epsilon$. Take δ to be the minimum distance from a to this set of rational numbers. Then, for $|p/q - a| < \delta$, we have $|T(x) - 0| = 1/q < \epsilon$. Hence, the limit is 0.

    ∎

As we saw in basic calculus, the algebra of limits can greatly simplify computations.

Theorem 2.9 (Algebra of Limits) *Let a be an accumulation point of the domains of both f and g. Suppose that f and g each have a finite limit at $x = a$ and let $c \in \mathbb{R}$. Then*

- $\lim_{x \to a} c\, f(x) = c \lim_{x \to a} f(x)$

- $\lim_{x \to a} f(x) \pm g(x) = \lim_{x \to a} f(x) \pm \lim_{x \to a} g(x)$

- $\lim_{x \to a} f(x) \cdot g(x) = \lim_{x \to a} f(x) \cdot \lim_{x \to a} g(x)$

- *if* $\lim_{x \to a} g(x) \neq 0$, *then* $\lim_{x \to a} \dfrac{f(x)}{g(x)} = \dfrac{\lim_{x \to a} f(x)}{\lim_{x \to a} g(x)}$

The proofs are straightforward and are left to the exercises.

The sandwich theorem is among the most useful tools for determining limits.

Theorem 2.10 (Sandwich Theorem) *Let g, f, and h be functions having a common domain D contained in a deleted neighborhood of a. Suppose $g(x) \leq f(x) \leq h(x)$ for all $x \in D$. If $\lim_{x \to a} g(x) = L = \lim_{x \to a} h(x)$, then $\lim_{x \to a} f(x) = L$.*

The proof of the sandwich theorem hinges on the inequality

$$L - \epsilon < g(x) \leq f(x) \leq h(x) < L + \epsilon$$

and is left to the exercises.

A limit analyzes the behavior of a function "next to" a point a. Combining that behavior with the actual value at the point a leads to the concept of continuity.

Continuity at a Point

Cauchy defined continuity in Chapter 2 of his 1821 text *Cours d'Analyse* (Cauchy, 1821, p. 43). Cauchy's definition is essentially the one used in modern calculus classes.

> The function $f(x)$ is continuous within given limits if between these limits an infinitely small increment i in the variable x always produces an infinitely small increment, $f(x + i) - f(x)$, in the function itself.

The modern form of Cauchy's definition is: f is continuous at a if and only if the limit of $f(x)$ as x approaches a is $f(a)$; i.e., $\lim_{x \to a} f(x) = f(a)$.

Extend Borzellino's function $f(x) = \sqrt{x} \sin(1/x)$ by defining $f(0) = 0$. Now ask the question, "Is f continuous at 0?" By Example 2.3, the answer is yes! *Explain why this question cannot be answered in an elementary calculus course.*

Is the function of Example 2.4, $g(x) = \sqrt{|x^2 - 1| - 1}$, continuous at 0? The limit is ill-posed as 0 is not an accumulation point of $\mathrm{dom}(g)$. Cauchy's definition is not applicable. A more general definition of continuity[2] used in the study of abstract metric spaces would identify 0 as a point of continuity. A simple addition fixes our definition.

Definition 2.5 *Let* $f : D \to \mathbb{R}$ *and let* $a \in D$. *Then* f *is* continuous *at* a *if and only if either* a *is an isolated point of* D *or* $\lim_{x \to a} f(x) = f(a)$. *Further,* f *is* continuous *on a set* $S \subseteq D$ *if* f *is continuous at each point of* S.

With our new definition of continuity, the function g above is continuous on its entire domain. *Verify this!*

Thomae's function provides a fascinating example that requires pondering and breaks the old heuristic "draw a graph without picking up the pencil."

■ **EXAMPLE 2.6   Thomae's Function and Continuity**

Recall $T : [0, 1] \to \mathbb{R}$ is defined by

$$T(x) = \begin{cases} \dfrac{1}{q} & x = \dfrac{p}{q} \in \mathbb{Q} \text{ in lowest terms} \\ 0 & x \notin \mathbb{Q} \end{cases}$$

See Figure 2.3.

We've seen for any $a \in [0, 1]$ that $\lim_{x \to a} T(x) = 0$. For a irrational, the limit equals the function value 0, while for any rational the limit does not equal the function value. We conclude that Thomae's function T is continuous on the irrationals in $[0, 1]$ and discontinuous on the rationals in $[0, 1]$.   ■

Thomae's function is a modification of Dirichlet's original example of a function discontinuous everywhere (with limits nowhere), $D(x) = 1$ if x is rational and 0 otherwise, the *characteristic function* of $\mathbb{Q}$. Not every mathematician was pleased by the difficulties that appeared in Dirichlet's, Thomae's, and others' examples. In 1893, Hermite wrote to Stieltjes,

> *I turn away with fright and horror from this lamentable plague of functions which do not have derivatives.*

Since continuity is based, in the main, on limits, the algebra of continuity follows directly from limit properties. Proofs are left to the exercises; remember to handle cases for isolated points of the domain.

Theorem 2.11 (Algebra of Continuity) *Let* $f : D \to \mathbb{R}$ *and* $g : D \to \mathbb{R}$ *be continuous at* a *and let* $c \in \mathbb{R}$. *Then*

1. *cf is continuous at a,*

---

[2]The definition used in *topology* is: f is continuous at a if and only if the inverse image of every open set containing $f(a)$ is open.

2. $f \pm g$ is continuous at a,

3. $f \cdot g$ is continuous at a,

4. if $g(a) \neq 0$, then f/g is continuous at a.

One of the most important results is that the composition of continuous functions is continuous. First, we prove a lemma showing that continuity commutes with limits.

Lemma 2.12 *If a function f is continuous at a and ϕ is a function such that $f(\phi(t))$ is defined and $\lim_{t \to t_0} \phi(t) = a$, then*

$$\lim_{t \to t_0} f(\phi(t)) = f\left(\lim_{t \to t_0} \phi(t)\right)$$

Proof: Let $\epsilon > 0$. Since f is continuous at a, we can find a $\delta_1 > 0$ so that, whenever $x \in \text{dom}(f)$ and $|x - a| < \delta_1$, we have $|f(x) - f(a)| < \epsilon$. Since $\lim_{t \to t_0} \phi(t) = a$, we can choose a $\delta > 0$ such that for $0 < |t - t_0| < \delta$ we have $|\phi(t) - a| < \delta_1$. Combine the inequalities to obtain the result. ∎

The theorem on composition is a direct consequence of the lemma. The proof is left to the exercises.

Theorem 2.13 (Continuity of Composition) *Let $f : A \to \mathbb{R}$ and $g : B \to \mathbb{R}$ where $f(A) \subseteq B$. Suppose that f is continuous at $x = a \in A$ and g is continuous at $x = f(a) \in B$; then $g \circ f$ is continuous at $x = a$.*

After investigating when functions are continuous, it's natural to ask, "When is a function *not* continuous?" Recall Table 1.1 of Chapter 1; we classified discontinuities into first kind, removable or jump, and second kind, infinite or oscillatory. A function that is either always increasing or always decreasing is called monotone. A monotone function cannot be "badly" discontinuous.

Theorem 2.14 *If f is monotone on $[a, b]$, then f can only have discontinuities of the first kind. Further, f can only have a countable number of discontinuities.*

Proof: Without loss of generality assume f is increasing. Suppose f is discontinuous at $\hat{x} \in (a, b)$. Then, because f is increasing, the one-sided limits of f both exist at $\hat{x}$ since, for example, if $x_1 < x_2 \in (\hat{x} - \delta, \hat{x})$, then $f(x_1) \leq f(x_2) \leq f(\hat{x})$. Hence $f\left((\hat{x} - \delta, \hat{x})\right)$ is bounded and has a supremum. This supremum is the limit from the left. *Show this!* The limit from the right is derived similarly. Therefore f has a jump discontinuity at $\hat{x}$. If $\hat{x} = a$, or b, then an analogous argument also classifies the discontinuity as a jump. ∎

An immediate corollary describes functions which have discontinuities of the second kind, either infinite or oscillatory.

Corollary 2.15 *If f has a discontinuity of the second kind at a, then f must change from increasing to deceasing in every neighborhood of a.*

Proof: Suppose not; then there is a neighborhood N on which f doesn't change. That is, f is monotone on N. By the previous theorem, the discontinuity must then be a jump, which is a contradiction.   ∎

Thomae's function is continuous on the irrationals and discontinuous on the rationals—can this be reversed? Can a function be continuous on the rationals and discontinuous on the irrationals? Somewhat surprisingly, the answer is no. We state the next theorem, deferring the proof.

Theorem 2.16 *The set of discontinuities of any real-valued function must be a countable union of closed sets.*

Unlike the rationals, the set of irrational numbers cannot be written as a countable union of closed sets. Hence, the irrationals cannot be the set of discontinuities of a function. Looking at this result from a different perspective, if f is discontinuous on the irrationals, then f is discontinuous on all of $\mathbb{R}$.

Continuity is a very strong condition, tying nearby behavior to the value of the function at a point. Being continuous on a closed interval is a very strong quality that it quite useful. If f is continuous on a closed interval, then the range can't have missing values.

Theorem 2.17 (Intermediate Value Theorem)   *If f is continuous on $[a, b]$ and k is between $f(a)$ and $f(b)$, then there exists a value $c \in (a, b)$ such that $f(c) = k$.*

Proof: Without loss of generality, we assume that $f(a) < k < f(b)$. Define the set $S = \{x \in [a, b] \mid f(x) < k\}$. Then $S \neq \emptyset$ since $a \in S$. Also S is bounded above by b. Therefore, by the completeness property of $\mathbb{R}$, $c = \sup S$ exists. We know that $a \leq c < b$. *Why?* Hence $c \in [a, b)$.

Suppose that $f(c) < k$. Then $\epsilon = k - f(c) > 0$. Since f is continuous at c, there is a $\delta > 0$ such that for all $x \in [a, b]$ with $c < x < c + \delta$ we have

$$|f(x) - f(c)| < \epsilon = k - f(c)$$

This inequality implies that $f(x) < (k - f(c)) + f(c) = k$. Thus $x \in S$, contradicting c being an upper bound of S. Thus $f(c) \not< k$.

A similar argument shows that $f(c) \not> k$. Therefore $f(c) = k$.   ∎

A partial converse to the intermediate value theorem tells us the intermediate value property is not enough, by itself, to guarantee continuity. We state the result without proof.

Theorem 2.18 *If $f : [a, b] \to [a, b]$ is one-to-one and has the intermediate value property, then f is continuous on $[a, b]$.*

Many real-life problems can be modeled by continuous functions on closed intervals. We optimize the problem by finding maxima and minima. Continuity on a closed interval ensures these optimizing values exist.

Theorem 2.19 (Extreme Value Theorem) *If $f : [a, b] \to \mathbb{R}$ is continuous, then*

1. *there exists $x_m \in [a, b]$ such that $f(x_m) = \min_{x \in [a,b]} f(x)$,*

2. *there exists $x_M \in [a, b]$ such that $f(x_M) = \max_{x \in [a,b]} f(x)$.*

Before the proof, we establish a lemma.

Lemma 2.20 *If f is continuous on a closed interval, then f is bounded on that interval.*

Proof: Suppose that f is unbounded. Consider the two intervals $[a, (a + b)/2]$ and $[(a + b)/2, b]$. Since f is unbounded, it must be unbounded on at least one of these subintervals. Choose $[a_1, b_1]$ to be a subinterval where f is unbounded. Divide $[a_1, b_1]$ into two equal subintervals, choosing $[a_2, b_2]$ to be one on which f is unbounded. Continue in this manner creating a sequence of intervals $[a_1, b_1] \supset [a_2, b_2] \supset [a_3, b_3] \supset \cdots$. The set $\{a_i\}$ is a bounded, increasing set of real numbers and so has a least upper bound α which is an element of each $[a_i, b_i]$. *Show this!* Since f is continuous at α, there is a neighborhood $N_\delta(\alpha)$ on which the values of f are within ϵ of $f(\alpha)$. But this neighborhood must contain an interval $[a_i, b_i]$ on which f is unbounded, contradicting the method of choosing $[a_i, b_i]$. *Why?* Therefore, f must be bounded on $[a, b]$.   ∎

Now the proof of the extreme value theorem.

Proof: Since f is continuous on $[a, b]$, then f is bounded and has a least upper bound, say M. Suppose there is no $x_M \in [a, b]$ where $f(x_M) = M$. Then $M - f(x) > 0$ and $g(x) = 1/(M - f(x))$ is continuous on $[a, b]$. Since M is the least upper bound, we can make $M - f(x)$ arbitrarily small, making g arbitrarily large, contradicting the boundedness lemma for g. Thus, there must be a value $x_M \in [a, b]$ where $f(x_M) = M$.

A similar argument shows that f takes on its minimum value.   ∎

Our last result is very famous. Weierstrass showed that a function that is continuous on a closed interval is "almost" a polynomial.

Theorem 2.21 (Weierstrass Approximation Theorem) *Suppose that $f : [a, b] \to \mathbb{R}$ is continuous. Then for any $\epsilon > 0$ there is a polynomial $p(x)$ approximating $f(x)$ so that*

$$\max_{x \in [a,b]} |f(x) - p(x)| < \epsilon$$

For a proof that constructs the polynomial approximants, see Rudin (1976).

Uniform Continuity

When a function is continuous on a set, often the choice of δ depends on the particular point in question along with ϵ. For some functions, a single value of δ may suffice for all points in the set. This uniformity of δ gives us *uniform continuity*.

Definition 2.6 (Uniform Continuity) *A function f is uniformly continuous on $E \subseteq \mathrm{dom}(f)$ if and only if for every $\epsilon > 0$ there is a $\delta > 0$ such that, whenever $x_1, x_2 \in E$ and $|x_1 - x_2| < \delta$, then $|f(x_1) - f(x_2)| < \epsilon$.*

In Figure 2.4, as x get closer to zero, δ must get smaller and smaller to be able to capture the function in the δ-ϵ box. In Figure 2.5, the same δ can be used throughout the interval. For nonuniform continuity, we write "$\delta = \delta(\epsilon, x)$" to emphasize δ's dependence on the value of x. For uniform continuity, we write "$\delta = \delta(\epsilon)$ *alone*" to emphasize δ's independence from x.

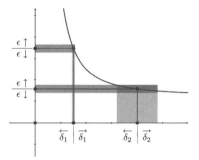

Figure 2.4   Nonuniform Continuity

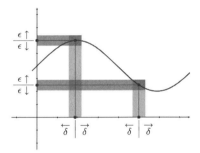

Figure 2.5   Uniform Continuity

A function that is continuous on a closed interval can always use one value of δ for a given ϵ no matter the x value.

Theorem 2.22 *If f is continuous on $[a, b]$, then f is uniformly continuous on $[a, b]$.*

The proof of this theorem is straightforward once we have the concept of *compact sets*. We will defer proving the result until then.

Continuity and uniform continuity tell us a great deal about a function. However, they only tell us that small changes in the independent variable produce small changes in the dependent variable. Can we find more information with deeper analysis? Yes. In the next section, we study the *rate* at which a function changes.

2.3   DIFFERENTIATION

The derivative as tangent has a long history that goes back to Euclid, Archimedes, and other ancients. The modern techniques for finding derivatives came from three perspectives. Descartes geometrically constructed tangent lines to various curves;

he later understood the usage for finding extrema. Fermat algebraically determined tangent lines, but his procedure included letting an increment be both nonzero when required and zero when needed—a logical conundrum. In the early 1600s, Roberval and Torricelli independently thought of tangents as arising from the direction vectors of a point moving along a curve [cf. Coolidge (1951)], giving a kinesthetic approach that would later be refined by Newton. We now blend these approaches in our classes.

Definition and Algebra of the Derivative

There are two forms of the standard definition of derivative.

Definition 2.7 *Let* $f : D \to \mathbb{R}$. *Suppose that* $a \in D$ *is an accumulation point of* D. *The* derivative of f *at* a *is defined by*

$$f'(a) = \lim_{x \to a} \frac{f(x) - f(a)}{x - a} = \lim_{h \to 0} \frac{f(a + h) - f(a)}{h}$$

if the limit exists. When $f'(a)$ *exists,* f *is said to be* differentiable at a.

While this definition is built on slopes, we need to keep in mind that the idea of "tangent line" is an interpretation, not a definition.

■ **EXAMPLE 2.7**

1. Let $f(x) = x^2 \sin(1/x)$ for $x \neq 0$ and $f(0) = 0$. Determine $f'(0)$. *Graph this function!*

   A simple calculation using the sandwich theorem finds $f'(0)$.

   $$f'(0) = \lim_{x \to 0} \frac{f(x) - 0}{x - 0} = \lim_{x \to 0} x \sin\left(\frac{1}{x}\right) = 0$$

2. Let $g(x) = x \sin(1/x)$ for $x \neq 0$ and $g(0) = 0$. Determine $g'(0)$. *Plot this one, too!*

   Another simple calculation shows $g'(0)$ does not exist.

   $$g'(0) = \lim_{x \to 0} \frac{g(x) - 0}{x - 0} = \lim_{x \to 0} \sin\left(\frac{1}{x}\right)$$

   Since $\lim_{x \to 0} \sin(1/x)$ does not exist, $g'(0)$ does not exist.

   ■

The sandwich theorem shows both functions in Example 2.7 are continuous; differentiability is a stronger condition than continuity. However, differentiability does imply continuity.

Theorem 2.23 *If f is differentiable at a, then f is continuous at a.*

Proof: Suppose f is differentiable at a. Then

$$\lim_{x \to a} f(x) - f(a) = \lim_{x \to a} \left[\frac{f(x) - f(a)}{x - a} \cdot (x - a) \right]$$

$$= \lim_{x \to a} \frac{f(x) - f(a)}{x - a} \cdot \lim_{x \to a} (x - a)$$

$$= f'(a) \cdot 0 = 0$$

Hence, $\lim_{x \to a} f(x) = f(a)$, which demonstrates that f is continuous at a.   ∎

■ **EXAMPLE 2.8**

Define the function h by

$$h(t) = \begin{cases} t - 1 & t \geq 0 \\ t^2 + t + 1 & t < 0 \end{cases}$$

Determine $h'(0)$.

It's quite tempting to look at

$$h'(t) = \begin{cases} 1 & t > 0 \\ 2t + 1 & t < 0 \end{cases}$$

as indicating that $h'(0) = 1$, but this is wrong. Looking at $\lim_{t \to 0-} h(t) = h(0-) = 1$ and $\lim_{t \to 0+} h(t) = h(0+) = -1$ shows that h is not continuous at $a = 0$. *Graph h!* The contrapositive of Theorem 2.23, if f is not continuous at a point, then f is not differentiable there, tells us $h'(0)$ cannot exist.   ∎

Our next task is to develop the algebra of derivatives. Just as with limits and continuity, the algebra of derivatives will extend the class of functions that we can easily differentiate.

Theorem 2.24 (Algebra of Differentiation) *Let f and g be differentiable at a and let $c \in \mathbb{R}$. Then cf, $f \pm g$, $f \cdot g$, and, if $g(a) \neq 0$, f/g are differentiable. Further*

1. $(c \cdot f)'(a) = c \cdot f'(a)$

2. $(f \pm g)'(a) = f'(a) \pm g'(a)$

3. $(f \cdot g)'(a) = f'(a) \cdot g(a) + f(a) \cdot g'(a)$

4. *if $g(a) \neq 0$, then* $\left(\dfrac{f}{g} \right)'(a) = \dfrac{f'(a) \cdot g(a) - f(a) \cdot g'(a)}{g^2(a)}$

Proof: We prove 3, leaving the others to the exercises.
Consider the difference quotient.

$$\frac{(fg)(x) - (fg)(a)}{x - a} = \frac{f(x)g(x) - f(a)g(a)}{x - a}$$

Add and subtract $f(a)g(x)$.

$$\frac{(fg)(x) - (fg)(a)}{x - a} = \frac{f(x)g(x) - f(a)g(x) + f(a)g(x) - f(a)g(a)}{x - a}$$

$$= \frac{f(x)g(x) - f(a)g(x)}{x - a} + \frac{f(a)g(x) - f(a)g(a)}{x - a}$$

$$= \frac{f(x) - f(a)}{x - a} g(x) + f(a) \frac{g(x) - g(a)}{x - a}$$

Take the limit as x approaches a. Then

$$\lim_{x \to a} \frac{(fg)(x) - (fg)(a)}{x - a} = \lim_{x \to a} \frac{f(x) - f(a)}{x - a} \cdot \lim_{x \to a} g(x) + f(a) \cdot \lim_{x \to a} \frac{g(x) - g(a)}{x - a}$$

$$= f'(a) \cdot g(a) + f(a) \cdot g'(a)$$

since f and g are differentiable at a and g must be continuous at a. ∎

It's interesting to note that Leibniz originally thought $(fg)' = f'g'$. He quickly discovered the error, becoming the first to find the product rule—we know this from reading his notebooks. Another interesting item is that computer algebra systems like Maple do not implement the quotient rule but apply the product rule to $f/g = f \cdot (g)^{-1}$. *Try it!*

With the algebra of derivatives, we can differentiate any polynomial or any rational function. To handle wider classes, we need the power of the chain rule.

Theorem 2.25 (The Chain Rule) *Let $f : A \to B$ and $g : B \to \mathbb{R}$. Suppose that f is differentiable at $x = a \in A$ and that g is differentiable at $x = b = f(a) \in B$. Then $g \circ f$ is differentiable at $x = a$ and*

$$(g \circ f)'(a) = g'(f(a)) \cdot f'(a)$$

Proof: Set $b = f(a)$. Since f and g are differentiable at a and b, there are functions $u(x)$ and $v(t)$, both going to zero as x and t go to a and b, respectively, such that

$$\frac{f(x) - f(a)}{x - a} = f'(a) + u(t)$$

$$\frac{g(t) - g(b)}{t - b} = g'(a) + v(t)$$

so that

$$f(x) - f(a) = (x - a)(f'(a) + u(x))$$
$$g(t) - g(b) = (t - b)(g'(b) + v(t))$$

Combining these formulas with $t = f(x)$ gives us

$$g(f(x)) - g(f(a)) = (f(x) - f(a)) \cdot (g'(b) + v(t))$$
$$= ((x - a) \cdot (f'(a) + u(x))) \cdot (g'(b) + v(f(x)))$$

If $x \neq a$, we may divide both sides by $x - a$. Then

$$\frac{g(f(x)) - g(f(a))}{x - a} = (f'(a) + u(x)) \cdot (g'(f(a)) + v(f(x)))$$

As x goes to a, we have $u(x)$ goes to zero. Also, as x goes to a, we have $f(x)$ goes to b, and so $v(f(x))$ goes to zero. Therefore,

$$\lim_{x \to a} \frac{g(f(x)) - g(f(a))}{x - a} = g'(f(a)) \cdot f'(a)$$

and the result is proven. ∎

It's tempting to try to use the expansion

$$\frac{g(f(x)) - g(f(a))}{x - a} = \frac{g(f(x)) - g(f(a))}{f(x) - f(a)} \cdot \frac{f(x) - f(a)}{x - a}$$

applying limits to both sides to prove the chain rule. Unfortunately, we can't guarantee $f(x) - f(a)$ is not zero, even though $x \neq a$. This problem can make the first term on the right side become undefined, invalidating the limit.

If f is differentiable and monotone, a formula for the derivative of f^{-1} comes directly from the chain rule.

Theorem 2.26 (Inverse Function Theorem) *Let $f : [a, b] \to \mathbb{R}$ be differentiable with $f'(x) \neq 0$ for any $x \in [a, b]$. Then*

- *f is one-to-one,*

- *f^{-1} is continuous on $f([a, b])$,*

- *f^{-1} is differentiable on $f([a, b])$,*

- *$(f^{-1})'(y) = \dfrac{1}{f'(x)}$ where $y = f(x)$.*

The proof is left to the reader with the hint that f composed with f^{-1} is the identity function.

Mean Value Theorems

The mean value theorems developed from observing a differentiable function's behavior near an extreme point. We know f has a *local maximum* (*local minimum*) at a when, on a neighborhood of a, f takes no larger (smaller) values. We call these points *local extrema*. Fermat made use of the derivative being zero at local extrema in his method to find maxima and minima.

Theorem 2.27 *Let f be defined on $[a, b]$ with a local extreme point at $c \in (a, b)$. If $f'(c)$ exists, then $f'(c) = 0$.*

Proof: Without loss of generality, assume c is a local maximum for f. Find a neighborhood $N_\delta(c) \subseteq (a, b)$ on which $f(x) \leq f(c)$. If $c - \delta < x < c$, then

$$\frac{f(x) - f(c)}{x - c} \geq 0$$

As $x \to c$, we have $f'(c) \geq 0$.
    If $c < x < c + \delta$, then

$$\frac{f(x) - f(c)}{x - c} \leq 0$$

As $x \to c$, we have $f'(c) \leq 0$. Hence $f'(c) = 0$.   ∎

Of course, an extreme value can occur at a point of nondifferentiability of a function; e.g., $|x|$ at $c = 0$. The converse does not hold either; this is easily seen by considering $f(x) = x^3$ at $c = 0$.
    If a function f is continuous on a closed interval, then it must have a maximum and minimum in the interval. If f is also differentiable, then the extreme values must have zero derivatives. If the function starts and ends at the same level but isn't constant, then the extreme values occur between the endpoints. This observation is Rolle's theorem. The proof is straightforward and left to the exercises.

Theorem 2.28 (Rolle's Theorem) *Suppose $f : [a, b] \to \mathbb{R}$ is*

1. *continuous on $[a, b]$,*

2. *differentiable on (a, b),*

3. *and $f(a) = f(b)$.*

Then there exists a point $c \in (a, b)$, such that $f'(c) = 0$.

The natural extension of Rolle's theorem "tilts" the function. Then we see that instead of f having a zero derivative, f's derivative must match the slope of the segment connecting the endpoints—the average or *mean* rate of change.

Theorem 2.29 (Lagrange's Mean Value Theorem) *Suppose* $f : [a, b] \to \mathbb{R}$ *is*

1. *continuous on* $[a, b]$ *and*

2. *differentiable on* (a, b).

Then there exists a point $c \in (a, b)$ *such that*

$$f'(c) = \frac{f(b) - f(a)}{b - a}$$

Proof: Define

$$\phi(x) = (f(x) - f(a)) - \frac{f(b) - f(a)}{b - a}(x - a)$$

Then ϕ is continuous on $[a, b]$, differentiable on (a, b), and $\phi(a) = 0 = \phi(b)$. Thus, apply Rolle's theorem to ϕ to obtain our result.   ■

On his way to proving the mean value theorem, Cauchy proved a result bounding the secant slopes with the derivative. An equivalent proposition, called the *race track principle* in Davis et al. (1994) and Hughes-Hallett et al. (2009) or the *speed limit law* in Ostebee & Zorn (2002), appears in today's calculus texts.

Theorem 2.30 *Let* f *be continuous on* $[a, b]$ *and differentiable on* (a, b). *Then*

$$\min_{x \in [a,b]} f'(x) \leq \frac{f(b) - f(a)}{b - a} \leq \max_{x \in [a,b]} f'(x)$$

While Cauchy gave an argument similar to the proof of Theorem 2.27, we can view the proposition as a direct corollary of Lagrange's mean value theorem since $\min f'(x) \leq f'(c) \leq \max f'(x)$ for every $c \in [a, b]$.

Cauchy gave a further generalization of Rolle's theorem extending to two functions. This theorem is also known as the *extended* or *generalized mean value theorem*.

Theorem 2.31 (Cauchy's Mean Value Theorem) *Suppose* $f, g : [a, b] \to \mathbb{R}$ *are*

1. *both continuous on* $[a, b]$ *and*

2. *both differentiable on* (a, b).

Then there exists a point $c \in (a, b)$, *such that*

$$f'(c)\left(g(b) - g(a)\right) = g'(c)\left(f(b) - f(a)\right)$$

Further, if $g(a) \neq g(b)$ *and* $g'(c) \neq 0$, *then*

$$\frac{f'(c)}{g'(c)} = \frac{f(b) - f(a)}{g(b) - g(a)}$$

The proof mimics that of Lagrange's theorem using a cleverly chosen function.

Proof: Define

$$\psi(x) = f(x)\left(g(b) - g(a)\right) - g(x)\left(f(b) - f(a)\right)$$

Apply Rolle's theorem to ψ to obtain the result.   ∎

Cauchy's mean value theorem has a nice geometric interpretation. Let's think of the two functions as a parametric reprsentation $\vec{r}(t) = [g(t), f(t)]$ for $t \in [a, b]$. Then the slope of a line tangent to the curve at $\vec{r}(c)$ is given by $f'(c)/g'(c)$. The slope of a line passing through the endpoints $\vec{r}(a)$ and $\vec{r}(b)$ is given by $m = (f(b) - f(a))/(g(b) - g(a))$. Cauchy's mean value theorem guarantees the existence of at least one point c where these slopes match, exactly paralleling the usual interpretation of the single-function mean value theorem. [Note: we've assumed $g'(c) \neq 0$ and $g(a) \neq g(b)$.]

■ **EXAMPLE 2.9**

Let $f(x) = \cos(x^2/5)$ and $g(x) = \sin(x/3)$ on $[0, 2\pi]$. Set $\vec{r}(t) = [g(t), f(t)]$.
In Figure 2.6, we see the curve $\vec{r}$ traveling from $P_0 = \vec{r}(0)$ to $P_1 = \vec{r}(2\pi)$. The hatched lines mark three tangents that are parallel to the line through P_0 and P_1. Points of tangency are labeled T. We need to solve the equation

$$\frac{f'(t)}{g'(t)} = \frac{f(2\pi) - f(0)}{g(2\pi) - g(0)}$$

or

$$-\frac{6t\sin(t^2/5)}{5\cos(t/3)} = \frac{2\sqrt{3}}{3}\left(\cos(4\pi^2/5) - 1\right)$$

to find values c for the tangent points T. We leave it to the exercises to find the three solutions in $(0, 2\pi)$ using a computer algebra system.   ■

Cauchy's mean value theorem can also be used to prove a result that was the subject of another priority controversy [see Burton (2007, p. 475)], the proposition we know as l'Hôpital's rule.

Theorem 2.32 (L'Hôpital's Rule) *Suppose functions f and g are continuous on $[a, b]$ and differentiable on (a, b). Further suppose that, for some $x_0 \in (a, b)$, $\lim_{x \to x_0} f(x) = 0$, $\lim_{x \to x_0} g(x) = 0$, and g is nonzero on a neighborhood of x_0. If $\lim_{x \to x_0} f'(x)/g'(x) = L$, then $\lim_{x \to x_0} f(x)/g(x) = L$.*

The proof is reasonably straightforward and is left to the reader.
A function may have a derivative at every point, but the derivative does not have to be continuous. For example, we've seen that $f(x) = x^2\sin(1/x)$ for $x \neq 0$ and $f(0) = 0$ is differentiable everywhere. However, f' is not continuous at zero. *Show this!* Remarkably though, derivatives do possess the intermediate value property.

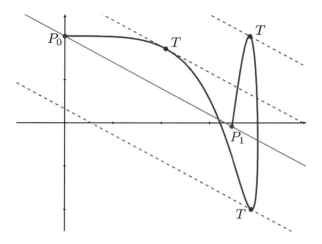

Figure 2.6   Cauchy's Mean Value Theorem Geometrically

Theorem 2.33 (Darboux's Intermediate Value Theorem for Derivatives) *If f is differentiable on $[a, b]$ and if k is some number between $f'(a)$ and $f'(b)$, then there exists a value $c \in (a, b)$ such that $f'(c) = k$.*

Proof: Assume without loss of generality that $f'(a) < k < f'(b)$. Define the function g by $g(x) = f(x) - k$. Then g is differentiable and $g'(x) = f'(x) - k$. Note that $g'(a) < 0$ and $g'(b) > 0$.

Since g is continuous on $[a, b]$, it must achieve a minimum, say at c. If $c = a$, then

$$\frac{g(x) - g(a)}{x - a} \geq 0$$

which implies that $g'(a) \geq 0$, a contradiction. Similarly $c \neq b$. Therefore $c \in (a, b)$. Since g has a minimum at c and is differentiable there, $g'(c) = 0$; i.e., $f'(c) = k$. ■

A corollary of Darboux's theorem describes what type of discontinuities derivatives can have.

Corollary 2.34 *A derivative cannot have simple discontinuities, only discontinuities of the second kind.*

This corollary tells us there are functions that cannot be the derivative of another function. For example, the Heaviside function $U(x) = 1$ if $x > 0$ and 0 otherwise has a jump discontinuity at $x = 0$. Therefore, U is not the derivative of any function *that includes 0 in its domain.*

Higher Derivatives and Taylor's Theorem

Brook Taylor published *Methodus Incrementorum Directa et Inversa* in 1715. In this text, he constructed the series expansion that we now call *Taylor series* even though, according to Stillwell (1989, p. 122), it "appears that Gregory used interpolation to discover Taylor's theorem 44 years before Taylor." Taylor also used polynomial interpolation with the points $[x_0, f(x_0)], [x_0+h, f(x_0+h)], [x_0+2h, f(x_0+2h)], \ldots$ to construct the series. The method is very clever but doesn't provide intuition. Then, in 1742, Maclaurin developed Taylor polynomials using "order of contact." That is, Maclaurin required the polynomial to match the function at successive derivatives. This technique is geometrically appealing: match the value, match the slope, match the concavity, etc. We'll follow Maclaurin's approach. See Hairer & Wanner (1996, p. 94) for an illustration of both methods.

Suppose that f has a derivative f' on an interval. If f' is differentiable, we call its derivative f''. For derivatives of order n, we use the notation $f^{(n)}$ or $d^n f / dx^n$. For $f^{(n)}$ to exist at a point x, $f^{(n-1)}$ must exist in a neighborhood of x and $f^{(n-1)}$ must be differentiable at x. Now, for $f^{(n-1)}$ to exist in a neighborhood of x, we need $f^{(n-2)}$ to be differentiable in that neighborhood, and so forth.

Suppose the function f is differentiable as many times as needed at $x = 0$. We wish to have a polynomial $p(x) = a_0 + a_1 x + a_2 x^2 + \cdots + a_n x^n$ that has *order of contact* n with f; that is, we want $p^{(k)}(0) = f^{(k)}(0)$ for $k = 0. \ldots . n$. (We use the convention that $f^{(0)} = f$.)

Substitute $x = 0$ to have $p(0) = a_0 = f(0)$. Substitute $x = 0$ into $p'(0) = a_1 = f'(0)$. Once more, substitute $x = 0$ into $p''(0) = 2a_2 = f''(0)$. In general, we see that $p^{(k)}(0) = k! \, a_k = f^{(k)}(0)$. These calculations lead us to take

$$p(x) = f(0) + f'(0)x + \frac{f''(0)}{2}x^2 + \cdots + \frac{f^{(n)}(0)}{n!}x^n$$

as our polynomial approximating f on some neighborhood of zero. To construct the polynomial about any other point a, simply apply the translation $x \to x - a$. We have just followed Maclaurin's approach to developing a Taylor series. Taylor's theorem is sometimes called the *Extended Law of the Mean*.

Theorem 2.35 (Taylor's Theorem) *If $f^{(n)}$ is continuous on the closed interval $[a, b]$ and if $f^{(n+1)}$ exists in (a, b), then for $x \in [a, b]$ there is a point $c \in [a, x)$ such that*

$$f(x) = \sum_{k=0}^{n} \frac{f^{(k)}(a)}{k!}(x - a)^k + \frac{f^{(n+1)}(c)}{(n + 1)!}(x - a)^{n+1}$$

The proof uses the same technique as Lagrange's mean value theorem: cleverly define a function and apply Rolle's theorem to it.

Proof: Let T_n be the nth-degree *Taylor polynomial centered at a*

$$T_n(x) = f(a) + \sum_{k=1}^{n} \frac{f^{(k)}(a)}{k!}(x - a)^k$$

Fix $x \in (a, b)$. Choose a constant K so that $f(x) - T_n(x) - K(x - a)^{n+1} = 0$. Define the ϕ by

$$\phi(t) = f(t) - T_n(t) - K(t - a)^{n+1}$$

Observe that ϕ is $n + 1$ times differentiable.

$$\phi'(t) = f'(t) - T_n'(t) - K(n + 1)(t - a)^n$$
$$\phi''(t) = f''(t) - T_n''(t) - K(n + 1)n\,(t - a)^{n-1}$$
$$\vdots$$
$$\phi^{(n)}(t) = f^{(n)}(t) - T_n^{(n)}(t) - K(n + 1)!\,(t - a)$$
$$\phi^{(n+1)}(t) = f^{(n+1)}(t) - K(n + 1)!$$

Since $\phi(a) = 0 = \phi(x)$, Rolle's theorem gives an $x_1 \in (a, x)$ where $\phi'(x_1) = 0$. Since $\phi'(a) = 0 = \phi'(x_1)$, Rolle's theorem gives an $x_2 \in (a, x_1)$ where $\phi''(x_2) = 0$.

Continue this process to obtain an $x_{n+1} \in (a, x_n)$ that gives $\phi^{(n+1)}(x_{n+1}) = 0$. But $\phi^{(n+1)}(x_{n+1}) = f^{(n+1)}(x_{n+1}) - K(n + 1)! = 0$ implies

$$K = \frac{f^{(n+1)}(x_{n+1})}{(n + 1)!}$$

Setting $c = x_{n+1}$ finishes the derivation.   ∎

The remainder term above can be used to estimate the error in approximating a function by a Taylor polynomial. We see

$$|f(x) - T_n(x)| \le \frac{(b - a)^{n+1}}{(n + 1)!} \cdot \max_{x \in [a,b]} \left| f^{(n+1)}(x) \right|$$

Another method of proof due to Cauchy produces the remainder term in integral form. Cauchy's remainder term is

$$R_n(x) = \frac{1}{n!} \int_a^x f^{(n+1)}(t)\,(x - t)^n\,dt$$

A Taylor series centered at zero is called a *Maclaurin series*. Recall we listed several important Maclaurin series in Example 1.13 in Chapter 1.

Even though Taylor and Maclaurin series are very useful, powerful tools, both for theory and practice, there are functions that cannot be approximated by these series. The canonical example is a function that is exceptionally "flat" at the origin.

■ **EXAMPLE 2.10**

Define Ψ by

$$\Psi(x) = \begin{cases} e^{-1/x^2} & x \ne 0 \\ 0 & \text{otherwise} \end{cases}$$

See Figure 2.7 for a graph of Ψ.

Since $\Psi^{(n)}(0) = 0$ for all n, (*Show this!*) the Maclaurin expansion of $\Psi(x)$ is 0. The Maclaurin series only matches the function at 0, nowhere else. ∎

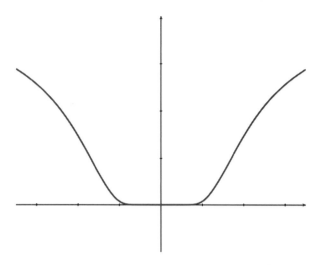

Figure 2.7   Graph of $\Psi(x)$

We have studied the derivative and looked at its properties and applications. As is typical in mathematics exposition, logical development does not follow historical development. We have defined the derivative and carefully built theorems to bring out its properties and uses. This sequence is almost opposite to 200 years of history. To quote Judith Grabiner (2004, p. 226) from "The Changing Concept of Change,"

> Fermat implicitly used it; Newton and Leibniz discovered it; Taylor, Euler, and Maclaurin developed it; Lagrange named it and characterized it; and only at the end of this long period of development did Cauchy and Weierstrass define it.

2.4 RIEMANN AND RIEMANN-STIELTJES INTEGRATION

Integration is the oldest area of calculus and analysis and yet one of the most modern. The usual motivation used to present the definite integral is as the "area under the curve." In *Journey Through Genius*, a book of the Great Theorems, Dunham's first chapter is "Hippocrates' Quadrature of the Lune" (Dunham, 1990). Hippocrates of Chios calculated the area bounded by a lune, the crescent area between two circular arcs, about 450 B.C. Three of Dunham's twelve great theorems concern area. But development didn't stop there. Jump forward to the beginning of the twentieth century: Lebesgue defined measure and a much more general form of integration that we'll study in Chapter 3. Currently, researchers also work to implement and extend the Risch algorithm, invented in 1968 by Robert Risch [see Risch (1969)], which determines whether an elementary antiderivative exists for a function and, if so, what that antiderivative must be.

While Newton and Leibniz are credited with being the first to understand the importance of the fundamental theorem tying integral to derivative, Cauchy was the first to give a formal, logical development of the definite integral—his definition still appears in calculus texts today as "left sums"; e.g. Ostebee & Zorn (2002, p. 373). Cauchy needed functions to be continuous to guarantee the existence of an integral. Riemann extended the definition so as to integrate a broader class of functions. As do most modern calculus texts, we'll start with Riemann's definition. Then we generalize following Stieltjes, taking the first steps towards Lebesgue theory. In Chapter 3, we will consider Lebesgue's more general integral. Read Bressoud's *A Radical Approach to Real Analysis* (2005) to see an exemplary presentation of the development of integration from the historical impetus of Fourier series. Read Burk's *A Garden of Integrals* (2007) for an excellent concise presentation of the development of integration from Cauchy to Riemann to Lebesgue and beyond.

The Riemann Integral

Riemann's theory of integration came from his study of the question, "When does a function have a convergent Fourier series?" Since Fourier coefficients are found by integration, the question became, "When can a function be integrated?" [See Hochkirchen (2004, Chapter 9).]

Let's start with Riemann sums.

Definition 2.8 *A partition of an interval $[a, b]$ is a collection of $n + 1$ points $P = \{x_k\}$ such that $a = x_0 < x_1 < x_2 < \cdots < x_n = b$. The* norm *of the partition P is*

$$\|P\| = \max\{\Delta x_1, \Delta x_2, \ldots, \Delta x_n\}$$

where $\Delta x_k = x_k - x_{k-1}$. A partition Q is a refinement *of P if $P \subseteq Q$. A* tagged partition *is a partition P with n tags c_k such that $c_k \in [x_{k-1}, x_k]$ for $k = 1, \ldots, n$.*

Choosing a partition sets the stage for a Riemann sum.

Definition 2.9 *Suppose that the function $f : [a, b] \to \mathbb{R}$ is bounded and that $P = \{x_0, \ldots, x_n\}$ with $\{c_k\}$ forms a tagged partition of $[a, b]$. Then the* Riemann sum *of f with respect to the tagged partition P is*

$$\mathcal{R}(P, f) = \sum_{k=1}^{n} f(c_k) \Delta x_k$$

Set $M_k = \sup_{x \in [x_{k-1}, x_k]} f(x)$ and $m_k = \inf_{x \in [x_{k-1}, x_k]} f(x)$. The upper *and* lower Darboux sums *of f relative to the tagged partition P are*

$$\mathcal{U}(P, f) = \sum_{k=1}^{n} M_k \Delta x_k \quad and \quad \mathcal{L}(P, f) = \sum_{k=1}^{n} m_k \Delta x_k$$

respectively.

Figure 2.8 shows one panel from the upper, lower, and Riemann sums.

Suppose f is bounded by $m \leq f(x) \leq M$. For any tagged partition P, it follows that

$$m(b-a) \leq \mathcal{L}(P,f) \leq \mathcal{R}(P,f) \leq \mathcal{U}(P,f) \leq M(b-a)$$

Show it! Hence, the upper and lower Riemann sums will be bounded and have a supremum and infimum. We use the sums to define *upper* and *lower Riemann integrals*. In 1875, Darboux was the first to use upper and lower integrals to develop the Riemann integral (Burk, 2007, p. 50).

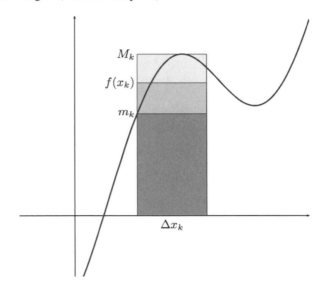

Figure 2.8 One Panel of a Riemann-Darboux Sum

Definition 2.10 *If* $f : [a,b] \to \mathbb{R}$ *is bounded, then define the* upper *and* lower Riemann integrals *to be*

$$\overline{\int_a^b} f(x)\,dx = \inf_P \mathcal{U}(P,f) \quad and \quad \underline{\int_a^b} f(x)\,dx = \sup_P \mathcal{L}(P,f)$$

respectively.

Now we are ready to define Riemann integrability in terms of the upper and lower integrals.

Definition 2.11 (Riemann Integrability) *A bounded function on the interval* $[a,b]$ *is* Riemann integrable *if and only if the upper and lower Riemann integrals are equal. We then write*

$$\underline{\int_a^b} f(x)\,dx = \overline{\int_a^b} f(x)\,dx = \int_a^b f(x)\,dx$$

Note: many authors emphasize independence from the variable of integration by using the notation $\int_a^b f$.

If f is bounded and monotone on $[a, b]$, then it is easy to determine the supremum and infimum on $[x_{k-1}, x_k]$ for a given partition. In that case, we can use the definition to determine the integral.

■ **EXAMPLE 2.11**

Find $\int_0^1 x^3 \, dx$.

We recognize that f is monotone increasing on $[0, 1]$; therefore

$$m_k = x_{k-1}^3 \qquad \text{and} \qquad M_k = x_k^3$$

Choose the partition P_n to be $P = \{0, 1/n, 2/n, \ldots, 1\}$. For each k, $\Delta x_k = 1/n$. Then

$$\mathcal{U}(P_n, f) = \sum_{k=1}^n \frac{k^3}{n^3} \Delta x = \frac{1}{4} \frac{(n+1)^2}{n^2}$$

and

$$\mathcal{L}(P_n, f) = \sum_{k=1}^n \frac{(k-1)^3}{n^3} \Delta x = \frac{1}{4} \frac{(n-1)^2}{n^2}$$

Figures 2.9 and 2.10 show upper and lower sums, respectively.

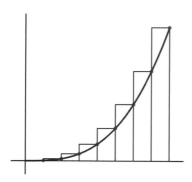

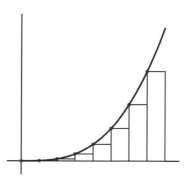

Figure 2.9   Upper Sums for x^3 **Figure 2.10**   Lower Sums for x^3

We see that both $\mathcal{U}(P_n, f)$ and $\mathcal{L}(P_n, f)$ go to $1/4$ as $n \to \infty$. Given arbitrary partitions P and Q, their common refinement is $P \cup Q$. It follows then that

$$\mathcal{L}(P_n, f) \leq \mathcal{L}(P_n \cup Q, f) \leq \mathcal{U}(P_n \cup Q, f) \leq \mathcal{U}(P_n, f)$$

Combining these observations gives us

$$\overline{\int}_0^1 x^3\,dx = \frac{1}{4} = \underline{\int}_0^1 x^3\,dx.$$

Thus $\int_0^1 x^3\,dx = 1/4$.   ∎

A function can be bounded but very discontinuous. Dirichlet's function, the character-istic function of the rationals, was called *the monster* by some of his contemporaries since it "broke everything."

■ **EXAMPLE 2.12**

Let D be Dirichlet's monster,

$$D(x) = \begin{cases} 1 & x \in \mathbb{Q} \\ 0 & \text{otherwise} \end{cases}$$

Determine $\int_0^1 D(x)\,dx$.

Let P be a partition of $[0, 1]$. Since every interval $[x_{k-1}, x_k]$ contains both rational and irrational numbers,

$$m_k = 0 \qquad \text{and} \qquad M_k = 1$$

whereupon

$$\mathcal{L}(P, f) = 0 \qquad \text{and} \qquad \mathcal{U}(P, f) = 1$$

Then, it's easy to see

$$\underline{\int}_0^1 D(x)\,dx = 0 \qquad \text{and} \qquad \overline{\int}_0^1 D(x)\,dx = 1$$

Since the upper and lower integrals are different, the Riemann integral doesn't exist.   ∎

The definition can be a difficult tool to use to find an integral or to prove one does not exist. We need more readily used criteria. Since a partition's upper and lower sums form bounds for any refinements sums, we look for a condition in that vein.

Theorem 2.36 *Let f be bounded on $[a, b]$. Then f is Riemann integrable over $[a, b]$ if and only if given an $\epsilon > 0$ there is a partition P such that $\mathcal{U}(P, f) - \mathcal{L}(P, f) < \epsilon$.*

Proof: First, suppose that f is Riemann integrable on $[a, b]$. Then the upper Riemann integral equals the lower. Call this common value A.

Let $\epsilon > 0$. Since $A - \epsilon/2 < \sup_P \mathcal{L}(P, f)$, there must be a partition P_1 for which

$$A - \epsilon/2 \le \mathcal{L}(P_1, f) \le \sup_P \mathcal{L}(P, f)$$

In the same fashion, there must be a partition P_2 so that

$$\inf_P \mathcal{L}(P, f) \leq \mathcal{U}(P_2, f) \leq A + \epsilon/2$$

Choose P to be a common refinement of P_1 and P_2. Then

$$\mathcal{U}(P, f) - \mathcal{L}(P, f) \leq \mathcal{U}(P_2, f) - \mathcal{L}(P_1, f)$$
$$\leq \mathcal{U}(P_2, f) - \inf_P \mathcal{U}(P, f) + \sup_P \mathcal{L}(P, f) - \mathcal{L}(P_1, f)$$
$$< \frac{\epsilon}{2} + \frac{\epsilon}{2} = \epsilon$$

For proving the other direction, suppose that f is bounded and $\epsilon > 0$ is given. Then there must be, by hypothesis, a partition P so that $\mathcal{U}(P, f) - \mathcal{L}(P, f) < \epsilon$. However

$$\mathcal{L}(P, f) \leq \underline{\int_a^b} f(x)\, dx \leq \overline{\int_a^b} f(x)\, dx \leq \mathcal{U}(P, f)$$

holds for any partition P. Therefore, we have

$$0 \leq \overline{\int_a^b} f(x)\, dx - \underline{\int_a^b} f(x)\, dx < \epsilon$$

Since $\epsilon > 0$ is arbitrary, the lower and upper Riemann integrals must be equal. Hence f is Riemann integrable. ∎

Let f be a continuous function on $[a, b]$. For any partition P, set $I_k = [x_{k-1}, x_k]$. Since

$$\lim_{\Delta x_k \to 0} \left[\sup_{x \in I_k} f(x) - \inf_{x \in I_k} f(x) \right] = 0$$

for continuous functions (*Show this!*), we expect continuous functions to be Riemann integrable. And they are.

Theorem 2.37 *If f is continuous on $[a, b]$, then f is Riemann integrable on $[a, b]$.*

Proof: Choose an arbitrary $\epsilon > 0$. Since f is continuous on the closed interval $[a, b]$, then f is uniformly continuous there. Thus, there must be a $\delta > 0$ so that whenever x_1 and $x_2 \in [a, b]$ satisfy $|x_1 - x_2| < \delta$, then $|f(x_1) - f(x_2)| < \epsilon/(b - a)$.

Select a partition P of $[a, b]$ with $\|P\| < \delta$. Then

$$\mathcal{U}(P, f) - \mathcal{L}(P, f) = \sum_{k=1}^n (M_k - m_k)\, \Delta x_k$$

By Theorem 2.19, the extreme value theorem, there are points $x_{1,k}$ and $x_{2,k} \in [x_{k-1}, x_k]$ for which $M_k = f(x_{k,1})$ and $m_k = f(x_{k,2})$. So

$$\mathcal{U}(P, f) - \mathcal{L}(P, f) = \sum_{k=1}^n (f(x_{k,1}) - f(x_{k,2})) \Delta x_k$$
$$\leq \sum_{k=1}^n \left(\frac{\epsilon}{b - a} \right) \Delta x_k = \epsilon$$

because each $|x_{k,1} - x_{k,2}| \leq x_k - x_{k-1} \leq \|P\| < \delta$. We have found a partition for which $\mathcal{U}(P, f) - \mathcal{L}(P, f) < \epsilon$; hence, f is Riemann integrable on $[a, b]$. ∎

Cauchy's definition of integration, based on what we now call left sums, requires a function to be continuous to be integrable. Can we weaken this condition and integrate a wider class of functions with Riemann's definition? Again, the answer is, "Yes."

Theorem 2.38 *If f is monotone on $[a, b]$, then f is Riemann integrable on $[a, b]$.*

Proof: Without loss of generality, we assume f is increasing on $[a, b]$. Let $\epsilon > 0$. Set $\Delta x = (b - a)/n$ and choose the equally spaced partition $P_n = \{a, a + \Delta x, a + 2\Delta x, \dots, a + (n-1)\Delta x, b\}$. Since f is increasing, $M_k = f(x_k)$ and $m_k = f(x_{k-1})$. Consider

$$\mathcal{U}(P_n, f) - \mathcal{L}(P_n, f) = \sum_{k=1}^{n} (f(x_k) - f(x_{k-1}))\Delta x$$

$$= \frac{b - a}{n} \cdot \sum_{k=1}^{n} (f(x_k) - f(x_{k-1}))$$

This sum is telescoping; thus

$$\mathcal{U}(P_n, f) - \mathcal{L}(P_n, f) = \frac{b - a}{n} \cdot (f(b) - f(a)).$$

The right side goes to 0 as $n \to \infty$, indicating that we can find an $n > 0$ so that $\mathcal{U}(P_n, f) - \mathcal{L}(P_n, f) < \epsilon$ for P_n. Thence, f is integrable on $[a, b]$. ∎

We could continue this line of thought to discover weaker classes of functions that are Riemann integrable, but we'll leave that for later. After we have some Lebesgue theory in our toolbox, we can describe which functions are Riemann integrable very concisely. At this point, however, we'll turn our attention to properties of integrable functions.

Properties of the Riemann Integral   Just as with limits, continuity, and differentiation, we consider algebraic properties of the Riemann integral. The following results parallel the earlier propositions. The proofs will be left to the exercises.

Theorem 2.39 (Algebra of the Riemann Integral) *Let f and g be Riemann integrable on $[a, b]$ and let $c \in \mathbb{R}$. Then*

1. *$f \pm g$ is integrable on $[a, b]$ and $\int_a^b (f \pm g)(x)\, dx = \int_a^b f(x)\, dx \pm \int_a^b g(x)\, dx$,*

2. *cf is Riemann integrable on $[a, b]$ and $\int_a^b cf(x)\, dx = c \int_a^b f(x)\, dx$,*

3. *if $c \in (a, b)$, then $\int_a^b f(x)\, dx = \int_a^c f(x)\, dx + \int_c^b f(x)\, dx$,*

4. *if $f(x) \leq g(x)$ on $[a, b]$, then $\int_a^b f(x)\, dx \leq \int_a^b g(x)\, dx$.*

A very useful property of integration allows us to reduce an integral by removing a factor. This theorem is due to Bonnet (1849), a nineteenth-century French mathematician.

Theorem 2.40 (Bonnet's Mean Value Theorem) *Suppose that f is continuous on $[a, b]$ and that g is nonnegative and integrable on $[a, b]$. Then there is a value $c \in (a, b)$ such that*

$$\int_a^b f(x)g(x)\,dx = f(c) \int_a^b g(x)\,dx$$

Proof: By hypothesis, f is continuous on a closed interval; therefore, f takes on maximum and minimum values, say M and m, respectively. Since g is nonnegative, we have $mg(x) \leq f(x)g(x) \leq Mg(x)$. Thus

$$m \int_a^b g(x)\,dx \leq \int_a^b f(x)g(x)\,dx \leq M \int_a^b g(x)\,dx$$

If $g(x) \equiv 0$, then we are done. *Why?* Assume $g(x) \not\equiv 0$. Then $\int_a^b g(x)\,dx > 0$ since g is nonnegative. So $m \leq \int_a^b f(x)g(x)\,dx / \int_a^b g(x)\,dx \leq M$. Apply the intermediate value theorem to the function f to obtain a point $c \in (a, b)$ so that $f(c) = \int_a^b f(x)g(x)\,dx / \int_a^b g(x)\,dx$, which finishes the proof.   ∎

Bonnet's theorem is also known as the *first mean value theorem for integrals*. The following corollary, with $g(x) = x$, is sometimes euphemistically called the "water in the bucket theorem."

Corollary 2.41 *If f is continuous on $[a, b]$, then there is a value $c \in (a, b)$ such that*

$$f(c)(b - a) = \int_a^b f(x)\,dx$$

The number $\bar{f} = \int_a^b f(x)\,dx/(b - a)$ is called the *average value* of f on $[a, b]$. Note there are three values between a and b that give $\bar{f}$ in Figure 2.11.

Using upper and lower integrals or sums based on partitions is a very difficult technique for evaluating integrals. The connection between differentiation and integration that Leibniz and Newton recognized as crucial provides the best method in a number of cases. Unfortunately, it can be very hard to find an antiderivative for a particular function. In fact, Richardson (1968) gave an example of a class of functions that are "undecidable"; i.e., it's not possible to determine whether antiderivatives exist or not. Nevertheless, the fundamental theorem is so powerful and useful that many beginning students see it as the definition of integral. That students would see the fundamental theorem as the definition comes as no surprise; books such as Granville et al. (1911) used the fundamental theorem as the definition and derived "area summation" as a property of the definite integral.

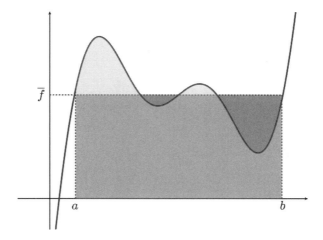

Figure 2.11   Mean Value Theorem for Integrals.

Theorem 2.42 (Fundamental Theorem of Calculus) *If F is differentiable on $[a, b]$ and $f = F'$ is integrable on $[a, b]$, then*

$$\int_a^b f(x)\, dx = F(b) - F(a)$$

Proof: Let P be a partition of $[a, b]$. Then

$$F(b) - F(a) = \sum_{k=1}^{n} F(x_k) - F(x_{k-1}) = \sum_{k=1}^{n} F'(t_k)\Delta x_k$$

where each t_k is found using the mean value theorem. Since $F' = f$ on $[a, b]$, we can rewrite the equation above as

$$F(b) - F(a) = \sum_{k=1}^{n} f(t_k)\Delta x_k$$

Given $\epsilon > 0$, we can choose the partition P so that

$$\left| \sum_{k=1}^{n} f(t_k)\Delta x_k - \int_a^b f(x)\, dx \right| < \epsilon$$

because f is integrable. Hence

$$\left| (F(b) - F(a)) - \int_a^b f(x)\, dx \right| < \epsilon$$

for any $\epsilon > 0$, and the theorem is proved.   ∎

Let's look at two classic examples for the fundamental theorem.

■ **EXAMPLE 2.13**

1. Let

$$f(x) = \begin{cases} \dfrac{x^3}{3} \sin\left(\dfrac{1}{x}\right) & x \neq 0 \\ 0 & x = 0 \end{cases}$$

From Exercise 2.31, we know that

$$f'(x) = \begin{cases} x^2 \sin\left(\dfrac{1}{x}\right) - \dfrac{x}{3} \cos\left(\dfrac{1}{x}\right) & x \neq 0 \\ 0 & x = 0 \end{cases}$$

which is bounded and also continuous at $x = 0$. See Figures 2.12 and 2.13. Since f' is continuous on $[-1, 1]$ (actually, *uniformly and absolutely continuous*), then it is integrable there. This function can be recovered from its derivative with the fundamental theorem.

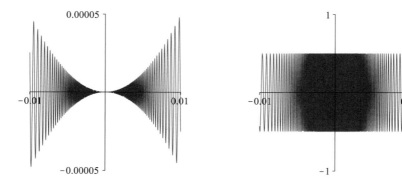

Figure 2.12   Graph of f of Example 2.13   **Figure 2.13**   Graph of f' of Example 2.13

2. Let

$$g(x) = \begin{cases} \dfrac{x^2}{2} \sin\left(\dfrac{1}{x^2}\right) & x \neq 0 \\ 0 & x = 0 \end{cases}$$

Again using Exercise 2.31, we have

$$g'(x) = \begin{cases} x \sin\left(\dfrac{1}{x^2}\right) - \dfrac{1}{x} \cos\left(\dfrac{1}{x^2}\right) & x \neq 0 \\ \text{DNE} & x = 0 \end{cases}$$

which is unbounded and not continuous at $x = 0$. See Figures 2.14 and 2.15. Since g and its derivative oscillate so quickly, the graphs appear to be solid regions. We know g' is unbounded, hence its integral does not exist on $[-1, 1]$. The fundamental theorem cannot be applied over any interval including 0.

■

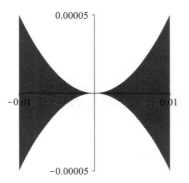
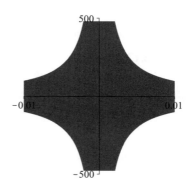

Figure 2.14   Graph of g of Example 2.13   **Figure 2.15**   Graph of g' of Example 2.13

There are many topics we could pursue: change of variables, integration by parts, improper integrals, and so forth. However, at this point, we will turn to generalizing Riemann's integral.

The Riemann-Stieltjes Integral

Stieltjes was a Dutch mathematician of the late 1800s who worked in nearly all branches of analysis and, like Riemann, unfortunately died young. Stieltjes received a prize for his work on analytic continued fractions, mostly undertaken at the University of Toulouse. It was Stieltjes' research in continued fractions that led him to generalize Riemann's formulation, developing what we now call the Riemann-Stieltjes integral.

So far, we have considered the x-axis to be a uniform horizontal line. Now, think of the axis as having mass or charge that is distributed nonuniformly along its length. These masses change the relative contribution of an interval to a Riemann sum. This view is what Stieltjes was considering: computing moments of masses distributed along the axis.

Definition 2.12 *Suppose α is monotonically increasing on $[a, b]$ and $f : [a, b] \to \mathbb{R}$ is bounded. Let $P = \{x_0, \ldots, x_n\}$ with $\{c_k\}$ form a tagged partition of $[a, b]$. Set $\Delta\alpha_k = \alpha(x_k) - \alpha(x_{k-1}) \geq 0$. Then the* **Riemann-Stieltjes sum** *of f with respect to the tagged partition P is*

$$\mathcal{R}(P, f, \alpha) = \sum_{k=1}^{n} f(c_k)\Delta\alpha_k$$

Set $M_k = \sup_{x \in [x_{k-1}, x_k]} f(x)$ and $m_k = \inf_{x \in [x_{k-1}, x_k]} f(x)$. The upper *and* lower Darboux-Stieltjes sums *of f relative to the tagged partition P are*

$$\mathcal{U}(P, f, \alpha) = \sum_{k=1}^{n} M_k \Delta \alpha_k \quad and \quad \mathcal{L}(P, f, \alpha) = \sum_{k=1}^{n} m_k \Delta \alpha_k$$

respectively.

As before, we use the sums to define *upper* and *lower Riemann-Stieltjes integrals*.

Definition 2.13 *If α is monotonically increasing on $[a, b]$ and $f : [a, b] \to \mathbb{R}$ is bounded, then define the* upper *and* lower Riemann-Stieltjes integrals *to be*

$$\overline{\int_a^b} f(x)\, d\alpha(x) = \inf_P \mathcal{U}(P, f, \alpha) \quad and \quad \underline{\int_a^b} f(x)\, d\alpha(x) = \sup_P \mathcal{L}(P, f, \alpha)$$

respectively.

A bounded function f on $[a, b]$ is Riemann-Stieltjes integrable *if and only if the upper and lower Riemann-Stieltjes integrals are equal. We then write*

$$\int_a^b f(x)\, d\alpha(x) = \overline{\int_a^b} f(x)\, d\alpha(x) = \underline{\int_a^b} f(x)\, d\alpha(x)$$

Note: authors also emphasize independence from the variable of integration in Riemann-Stieltjes integrals by using the notation $\int_a^b f\, d\alpha$.

When does a Riemann-Stieltjes integral exist? One answer matches a condition for Riemann integrals.

Theorem 2.43 *A function f is Riemann-Stieltjes integrable with respect to α on $[a, b]$ if and only if for every $\epsilon > 0$ there is a partition P such that $\mathcal{U}(P, f, \alpha) - \mathcal{L}(P, f, \alpha) < \epsilon$.*

Proof: Given any partition P, it follows from the definition that

$$\mathcal{L}(P, f, \alpha) \leq \underline{\int_a^b} f\, d\alpha \leq \overline{\int_a^b} f\, d\alpha \leq \mathcal{U}(P, f, \alpha)$$

Thus, if $\mathcal{U}(P, f, \alpha) - \mathcal{L}(P, f, \alpha) < \epsilon$ for an arbitrary $\epsilon > 0$, then $\underline{\int_a^b} f\, d\alpha = \overline{\int_a^b} f\, d\alpha$ and f is Riemann-Stieltjes integrable with respect to α.

For the other direction, suppose that f is Riemann-Stieltjes integrable with respect to α on $[a, b]$. Then there are partitions P_1 and P_2 for which

$$\mathcal{U}(P_1, f, \alpha) - \int_a^b f\, d\alpha < \frac{\epsilon}{2} \quad and \quad \int_a^b f\, d\alpha - \mathcal{L}(P_2, f, \alpha) < \frac{\epsilon}{2}$$

Set $P = P_1 \cup P_2$. We see that

$$\mathcal{U}(P, f, \alpha) \leq \mathcal{U}(P_1, f, \alpha) < \int_a^b f\, d\alpha + \frac{\epsilon}{2} < \mathcal{L}(P_2, f, \alpha) + \epsilon < \mathcal{L}(P, f, \alpha) + \epsilon$$

That is,

$$\mathcal{U}(P, f, \alpha) - \mathcal{L}(P, f, \alpha) < \epsilon$$

∎

Let's look at a pair of integrals that will help us to understand the interplay between f and α.

■ **EXAMPLE 2.14**

1. Let α be the unit step function $\alpha(x) = 1$ if $x > 0$ and $\alpha(x) = 0$ otherwise. Let f be bounded on $[-1, 1]$ and continuous at $x = 0$. Then $\int_{-1}^{1} f(x)\, d\alpha(x) = f(0)$. Since P is a partition, 0 is in exactly one subinterval $[x_{\hat{k}-1}, x_{\hat{k}}]$. Then $\Delta\alpha_{\hat{k}} = 1$; all other $\Delta\alpha_k = 0$. Therefore

$$\mathcal{U}(P, f, \alpha) = M_{\hat{k}} \cdot \Delta\alpha_{\hat{k}} = M_{\hat{k}} \quad \text{and} \quad \mathcal{L}(P, f, \alpha) = m_{\hat{k}} \cdot \Delta\alpha_{\hat{k}} = m_{\hat{k}}$$

By the continuity of f at $x = 0$, we see

$$\overline{\int}_{-1}^{1} f\, d\alpha = \lim_{\|P\| \to 0} M_{\hat{k}} = f(0) \quad \text{and} \quad \underline{\int}_{-1}^{1} f\, d\alpha = \lim_{\|P\| \to 0} m_{\hat{k}} = f(0)$$

The upper and lower integrals are equal; therefore $\int_{-1}^{1} f\, d\alpha = f(0)$.

2. Let α be the unit step function. Show that $\int_{-1}^{1} \alpha(x)\, d\alpha(x)$ does not exist.

As before, any Riemann-Stieltjes sum has a single nonzero term with $M_{\hat{k}} = 1$ and $m_{\hat{k}} = 0$. Now, however, the upper and lower integrals are not the same.

$$\overline{\int}_{-1}^{1} \alpha\, d\alpha = \lim_{\|P\| \to 0} 1 = 1 \quad \text{and} \quad \underline{\int}_{-1}^{1} \alpha\, d\alpha = \lim_{\|P\| \to 0} 0 = 0$$

Thus, the integral $\int_{-1}^{1} \alpha(x)\, d\alpha(x)$ does not exist.

∎

These examples show that if α and f have a common discontinuity, the integral $\int_{a}^{b} f\, d\alpha$ doesn't exist, but if f is continuous at α's discontinuity (or vice versa), the integral can exist. If α is differentiable, we can say quite a bit more.

Theorem 2.44 *If α is differentiable (and monotone increasing) on $[a, b]$, and α' and f are Riemann integrable on $[a, b]$, then f is Riemann-Stieltjes integrable with respect to α on $[a, b]$, and*

$$\int_{a}^{b} f(x)\, d\alpha(x) = \int_{a}^{b} f(x)\, \alpha'(x)\, dx$$

Proof: Let an $\epsilon > 0$ be given and let $M > 0$ be a bound for $|f|$. Since α' is Riemann integrable, there is a tagged partition P of $[a, b]$ such that

$$\mathcal{U}(P, \alpha') - \mathcal{L}(P, \alpha') < \frac{\epsilon}{M}$$

By the mean value theorem, we can find n points t_k for which $\Delta\alpha_k = \alpha'(t_k)\Delta x_k$. Since $\mathcal{U}-\mathcal{L} < \epsilon/M$, if we choose any other set of tag points $s_k \in [x_{k-1}, x_k]$, then

$$\sum_{k=1}^{n} |\alpha'(s_k) - \alpha'(t_k)|\, \Delta x_k < \frac{\epsilon}{M}$$

By our choice of the points t_k, we see that

$$\sum_{k=1}^{n} f(s_k)\Delta\alpha_k = \sum_{k=1}^{n} f(s_k)\alpha'(t_k)\Delta x_k$$

Hence

$$\left| \sum_{k=1}^{n} f(s_k)\Delta\alpha_k - \sum_{k=1}^{n} f(s_k)\alpha'(s_k)\Delta x_k \right|$$

$$= \left| \sum_{k=1}^{n} f(s_k)\alpha'(t_k)\Delta x_k - \sum_{k=1}^{n} f(s_k)\alpha'(s_k)\Delta x_k \right|$$

$$= \left| \sum_{k=1}^{n} f(s_k)\left(\alpha'(t_k) - \alpha'(s_k)\right)\Delta x_k \right| < \epsilon$$

Since s_k was arbitrarily chosen, we then have

$$\mathcal{U}(P, f, \alpha) \leq \mathcal{U}(P, f\alpha') + \epsilon$$

In the same fashion,

$$\mathcal{U}(P, f\alpha') \leq \mathcal{U}(P, f, \alpha) + \epsilon$$

Therefore,

$$|\mathcal{U}(P, f\alpha') - \mathcal{U}(P, f, \alpha)| < \epsilon$$

Since this relation is true of any refinement of P, we have

$$\left| \overline{\int_a^b} f d\alpha - \overline{\int_a^b} f(x)\,\alpha'(x)\,dx \right| < \epsilon$$

The upper integrals must be equal as $\epsilon > 0$ was chosen arbitrarily. An analogous argument establishes the equality of the lower integrals. The integrals must all be equal because $f\alpha'$ is Riemann integrable on $[a, b]$. Hence,

$$\int_a^b f(x)\,d\alpha(x) = \int_a^b f(x)\,\alpha'(x)\,dx$$

∎

We can now handle a Riemann-Stieltjes integral with an α that is differentiable or has jump discontinuities. Algebra will extend our reach.

Properties of the Riemann-Stieltjes Integral The Riemann-Stieltjes integral also has a set of useful algebraic properties. We leave the proofs to the exercises.

Theorem 2.45 (Algebra of the Riemann-Stieltjes Integral) *Suppose f and g are Riemann-Stieltjes integrable with respect to both α and β on $[a, b]$. Let $c \in \mathbb{R}$. Then*

1. *cf is integrable with respect to α and*

$$\int_a^b cf \, d\alpha = c \int_a^b f \, d\alpha$$

2. *$f + g$ is integrable and*

$$\int_a^b f + g \, d\alpha = \int_a^b f \, d\alpha + \int_a^b g \, d\alpha$$

3. *if $c > 0$, then f is integrable with respect to $c\alpha$ and*

$$\int_a^b f \, d(c\alpha) = c \int_a^b f \, d\alpha$$

4. *f is integrable with respect to $\alpha + \beta$ and*

$$\int_a^b f \, d(\alpha + \beta) = \int_a^b f \, d\alpha + \int_a^b f \, d\beta$$

5. *if $c \in (a, b)$, then*

$$\int_a^b f \, d\alpha = \int_a^c f \, d\alpha + \int_c^b f \, d\alpha$$

6. *if $f(x) \le g(x)$ on $[a, b]$, then*

$$\int_a^b f \, d\alpha \le \int_a^b g \, d\alpha$$

The algebraic properties allow us to do many things: separate functions via linearity of the integral, decompose α into a sum of a function and a set of jumps, and so forth.

■ **EXAMPLE 2.15**

Let $f(x) = x$ and $\alpha(x) = x^2 + \lfloor x \rfloor$ on $[0, 2]$. Calculate

$$\int_0^2 f \, d\alpha$$

Using the algebraic properties of Riemann-Stieltjes integrals, we have

$$\int_0^2 x \, d(x^2 + \lfloor x \rfloor) = \int_0^2 x \, d(x^2) + \int_0^2 x \, d(\lfloor x \rfloor)$$

Since $(x^2)' = 2x$,

$$\int_0^2 x \, d(x^2 + \lfloor x \rfloor) = \int_0^2 2x^2 \, d(x) + \int_0^2 x \, d(\lfloor x \rfloor)$$

Our last simplification is to replace the second integral with the sum over the jumps of the floor function in $[0, 2]$; i.e., at $j = 1$ and 2.

$$\int_0^2 x \, d(x^2 + \lfloor x \rfloor) = \int_0^2 2x^2 \, dx + \sum_{j=1}^2 j$$

A simple evaluation yields

$$\int_0^2 x \, d(x^2 + \lfloor x \rfloor) = \frac{25}{3}$$

Verify this!   ∎

To go further than algebraic combinations, applying a continuous function to an integrable function yields an integrable function.

Theorem 2.46 *Suppose f is Riemann-Stieltjes integrable with respect to α on $[a, b]$ and is bounded by $m \le f(x) \le M$. Further, suppose g is continuous on $[m, M]$. Then $g \circ f$ is Riemann-Stieltjes integrable with respect to α on $[a, b]$.*

Proof: Let $\epsilon > 0$ be given. Since g is uniformly continuous on $[m, M]$, there is a $\delta > 0$ such that $\delta < \epsilon$ and $|g(s) - (t)| < \epsilon$ whenever $|s - t| < \delta$ for $s, t \in [m, M]$.
Since f is integrable, there is a partition P for which

$$\mathcal{U}(P, f, \alpha) - \mathcal{L}(P, f, \alpha) < \delta^2$$

For $k = 1, \ldots, n$, define

$$M_k = \sup_{[x_{k-1}, x_k]} f(x) \qquad \text{and} \qquad m_k = \inf_{[x_{k-1}, x_k]} f(x)$$

$$\widehat{M_k} = \sup_{[x_{k-1}, x_k]} (g \circ f)(x) \qquad \text{and} \qquad \widehat{m_k} = \inf_{[x_{k-1}, x_k]} (g \circ f)(x)$$

and

$$K = \sup_{[m, M]} g(x)$$

Separate the indices k into two disjoint sets: put $k \in A$ if $M_k - m_k < \delta$, otherwise put $k \in B$.
If $k \in A$, then for any $s, t \in [x_{k-1}, x_k]$, it follows that $|f(s) - f(t)| < \delta$ and hence $|(g \circ f)(s) - (g \circ f)(t)| < \epsilon$.
If $k \in B$, then $\delta \le \widehat{M_k} - \widehat{m_k} \le 2K$. Since $\mathcal{U} - \mathcal{L} < \delta^2$, we see that

$$\delta \sum_{k \in B} \Delta \alpha_k \le \sum_{k \in B} (M_k - m_k) \Delta \alpha_k \le \mathcal{U}(P, f, \alpha) - \mathcal{L}(P, f, \alpha) < \delta^2$$

Thus $\sum_{k \in B} \Delta \alpha_k < \delta$.

Put these inequalities together to yield

$$\mathcal{U}(P, g \circ f, \alpha) - \mathcal{L}(P, g \circ f, \alpha) = \sum_{k \in A} (\widehat{M_k} - \widehat{m_k}) \Delta \alpha_k + \sum_{k \in B} (\widehat{M_k} - \widehat{m_k}) \Delta \alpha_k$$
$$\leq \sum_{k \in A} (\epsilon) \Delta \alpha_k + \sum_{k \in B} (2K) \Delta \alpha_k$$
$$\leq \epsilon \sum_{k \in A} \Delta \alpha_k + 2K \sum_{k \in B} \Delta \alpha_k$$

Extend the first sum to all k and replace the second by δ. Then

$$\leq \epsilon \sum_{k=1}^{n} \Delta \alpha_k + 2K\delta$$

Recall that δ was chosen to be less than ϵ; hence

$$\mathcal{U}(P, g \circ f, \alpha) - \mathcal{L}(P, g \circ f, \alpha) < \epsilon(\alpha(b) - \alpha(a) + 2K)$$

Since ϵ was arbitrary, we have that $g \circ f$ is Riemann-Stieltjes integrable with respect to α, and the result is proven. ∎

Our last result of the section is a direct application of the previous theorem and the algebraic properties of Riemann-Stieltjes integrals. The proof is straightforward and is left to the exercises.

Theorem 2.47 *Suppose f is Riemann-Stieltjes integrable with respect to α on $[a, b]$. Then*

1. *if $m \leq f(x) \leq M$ on $[a, b]$, then*

$$m \cdot (\alpha(b) - \alpha(a)) \leq \int_a^b f \, d\alpha \leq M \cdot (\alpha(b) - \alpha(a))$$

2. *$|f|$ is Riemann-Stieltjes integrable with respect to α on $[a, b]$ and*

$$\left| \int_a^b f \, d\alpha \right| \leq \int_a^b |f| \, d\alpha$$

We've investigated a number of the properties of Riemann and Riemann-Stieltjes integration, but we still haven't completely answered the question of which functions are integrable. The easiest method we can use to answer this question requires Lebesgue measure, so we'll postpone the answer until Chapter 3. As we go forward, keep thinking about the question.

2.5 SEQUENCES, SERIES, AND CONVERGENCE TESTS

Sequences and series have appeared in mathematics since antiquity: from Zeno's paradox, Eudoxes' method of exhaustion, Archimedes' quadratures, Fibonacci's rabbits, Suiseth and Oresme's calculations of accelerations, Euler's solution of the Basel problem, to Ramanujan's series for π. We had to wait, however, for Cauchy for a precise definition of convergence and for tests to determine convergence. Bolzano had given a definition earlier in 1816, but his work was mostly unknown. See also Hairer & Wanner (1996, Section III.1).

Sequences of Constants

We start our investigations by moving from an informal list of numbers to a function on the natural numbers.

Definition 2.14 *A real-valued sequence is a function $a : \mathbb{N} \to \mathbb{R}$. Sequence terms are denoted by $a(n) = a_n$.*
   Let $\phi : \mathbb{N} \to \mathbb{N}$ be a strictly increasing function. The composition $a \circ \phi$ forms a subsequence *and is denoted by $a(\phi(k)) = a_{n_k}$ where $n_k = \phi(k)$.*

Cauchy's definition of convergence matches his definition for the limit of a function, again making the idea of "close" precise.

Definition 2.15 *A sequence* converges *to a number L, written*

$$\lim_{n \to \infty} a_n = L$$

if and only if for every $\epsilon > 0$ there is an $N \in \mathbb{N}$ such that if $n > N$ then $|a_n - L| < \epsilon$.

In words, if there is an N depending on ϵ so that we can capture the series in the ϵ-strip about the proposed limit L for all $n > N$, then the limit of the sequence is L, as shown in Figure 2.16. A sequence that doesn't converge is said to *diverge*.
   Let's consider an example of a limit proof.

■ **EXAMPLE 2.16**

   Prove
$$\lim_{n \to \infty} \frac{2n^2 + n + 2}{n^2 + 1} = 2$$
   Let $\epsilon > 0$ be given. We need to find an $N \in \mathbb{N}$ so that $n > N$ guarantees

$$\left| \frac{2n^2 + n + 2}{n^2 + 1} - 2 \right| = \frac{n}{n^2 + 1} < \epsilon$$

   A little bit of algebra simplifies the problem. Choose $N > 1/\epsilon$ by the Archimedean property of the real numbers. Now, if $n > N$, then

$$n + \frac{1}{n} > n > N > \frac{1}{\epsilon}$$

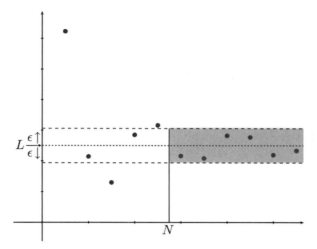

Figure 2.16   The N-ϵ Strip of Convergence

Since $n + 1/n = (n^2 + 1)/n$, we have, for $n > N$, the inequality $(n^2 + 1)/n > 1/\epsilon$. Invert to obtain

$$\frac{n}{n^2 + 1} = \left| \frac{2n^2 + n + 2}{n^2 + 1} - 2 \right| < \epsilon$$

Thus, the limit holds.   ∎

Note we must use an arbitrary ϵ; showing the inequality true for a particular value, such as $\epsilon = 0.01$, doesn't prove the limit exists. See Exercise 2.49 for an example.
    Next we examine whether a sequence can have several limits.

Theorem 2.48 *The limit of a convergent sequence is unique.*

Proof: Let the sequence $\{a_n\}$ converge to both L and M, and let $\epsilon > 0$.
    Since $a_n \to L$, there is an $N_1 \in \mathbb{N}$ such that

$$|a_n - L| < \frac{\epsilon}{2} \qquad n > N_1$$

Since $a_n \to M$, there is an $N_2 \in \mathbb{N}$ such that

$$|a_n - M| < \frac{\epsilon}{2} \qquad n > N_2$$

Set $N = \max\{N_1, N_2\}$ and let $n > N$. Then

$$|L - M| \leq | L - a_n| + |a_n - M| < \frac{\epsilon}{2} + \frac{\epsilon}{2} = \epsilon$$

Since $\epsilon > 0$ is arbitrary, $L = M$.   ∎

A function is bounded when its range is a bounded set. A sequence is a function, so a sequence $\{a_n\}$ is bounded when there is a constant M such that $|a_n| \leq M$ for all n. If a sequence converges to a limit L, then it is eventually bounded by $L + \epsilon$. Since there are finitely many terms before the "capturing ϵ-strip," the entire sequence is bounded.

Theorem 2.49 *A convergent sequence is bounded.*

Proof: Let the sequence $\{a_n\}$ converge to L and let $\epsilon = 1$. There is an $N \in \mathbb{N}$ such that $|a_n - L| < 1$ for all $n > N$. The triangle inequality then gives

$$|a_n| \leq | a_n - L| + |L| < 1 + |L|$$

for all $n > N$. The required bound M can be found from the maximum of the finite set

$$M = \max\{|a_1|, |a_2|, \ldots, |a_N|, 1 + |L|\}$$

By construction, $|a_n| < M$ for all n.   ∎

Is the converse true? That is, if a sequence is bounded, must it converge? No! Look at $a_n = \sin(n\pi/2)$.

When will a set of real numbers contain a convergent sequence? A first answer is if the set has an accumulation point.

Theorem 2.50 *If a is an accumulation point of $A \subseteq \mathbb{R}$, then there is a sequence $\{a_n\} \subseteq A$ converging to a.*

Proof: Let $\epsilon > 0$. For each $n \in \mathbb{N}$, choose a point $a_n \in N_{1/n}(p)$. *How do we know such points exist?* Choose N so large that $\epsilon > 1/N$; then for $n > N$, we have $|a_n - a| < 1/n < \epsilon$. Thus $a_n \to a$.   ∎

We turn to the algebra of limits of sequences to enhance our toolbox.

Theorem 2.51 *Let $\{a_n\}$ and $\{b_n\}$ be convergent sequences and let $c \in \mathbb{R}$. Then*

1. $\displaystyle\lim_{n\to\infty} (c\,a_n) = c \lim_{n\to\infty} a_n.$

2. $\displaystyle\lim_{n\to\infty} (a_n \pm b_n) = \lim_{n\to\infty} a_n \pm \lim_{n\to\infty} b_n.$

3. $\displaystyle\lim_{n\to\infty} (a_n \cdot b_n) = \lim_{n\to\infty} a_n \cdot \lim_{n\to\infty} b_n.$

4. $\displaystyle\lim_{n\to\infty} \frac{a_n}{b_n} = \frac{\lim\limits_{n\to\infty} a_n}{\lim\limits_{n\to\infty} b_n}$ *if* $\displaystyle\lim_{n\to\infty} b_n \neq 0.$

5. $\displaystyle\lim_{n\to\infty} a_n^k = \left[\lim_{n\to\infty} a_n\right]^k$ *for all $k \in \mathbb{N}$.*

6. *if $a_n \leq b_n$ for all $n > m$ for some $m \in \mathbb{N}$, then* $\displaystyle\lim_{n\to\infty} a_n \leq \lim_{n\to\infty} b_n.$

Proof: We prove part 3, leaving the other parts as an exercise.

Let $\epsilon > 0$. Since $\{b_n\}$ is convergent, the sequence is bounded in absolute value, say by $M > 0$. Let $a_n \to A$ and $b_n \to B$. There is an N_1 such that for all $n > N_1$

$$|a_n - A| < \frac{\epsilon}{2M}$$

There is an N_2 such that for all $n > N_2$

$$|b_n - B| < \frac{\epsilon}{2|A| + 1}$$

Set $N = \max\{N_1, N_2\}$ and let $n > N$. Then

$$|a_n b_n - AB| \leq |a_n b_n - b_n A| + |b_n A - AB|$$
$$\leq |b_n| \cdot |a_n - A| + |A| \cdot |b_n - B|$$

Apply the inequalities above to $|a_n - A|$ and $|b_n - B|$, noting $|b_n| < M$, to see

$$|a_n b_n - AB| < M \cdot \frac{\epsilon}{2M} + |A| \cdot \frac{\epsilon}{2|A| + 1}$$

Since $|A|/(2|A| + 1) < 1/2$, then

$$|a_n b_n - AB| < \frac{\epsilon}{2} + \frac{\epsilon}{2} = \epsilon$$

thus proving part 3. ∎

We should pause to reflect on the similarity of our set of "algebra theorems" to those we've already encountered.

Theorem 2.52 (Sandwich Theorem for Sequences) *Suppose that $\{a_n\}$, $\{b_n\}$, and $\{c_n\}$ are three sequences such that $a_n \leq b_n \leq c_n$ for all $n > m$ for some $m \in \mathbb{N}$. If $\lim a_n = \lim c_n = B$, then $\lim b_n = B$.*

Proof: Let $\epsilon > 0$. Since $a_n \to B$, there is an $N_1 \in \mathbb{N}$ so that $|a_n - B| < \epsilon$ for any $n > N_1$. In particular, for $n > N_1$, we have $B - \epsilon < a_n$. Similary, since $c_n \to B$, there is an $N_2 \in \mathbb{N}$ so that $|c_n - B| < \epsilon$ for any $n > N_2$. In particular, for $n > N_2$, we have $c_n < B + \epsilon$. Thus, for $n > N = \max\{N_1, N_2\}$, we have

$$B - \epsilon < a_n \leq b_n \leq c_n < B + \epsilon$$

which implies that $|b_n - B| < \epsilon$. Hence $b_n \to B$. ∎

So far, our methods have been aimed at proving convergence to a known value. Can we determine if a sequence converges without pre-knowledge of the limit? Yes! We begin with increasing sequences that are bounded above.

Theorem 2.53 *If $\{a_n\}$ is a bounded, increasing sequence, then $\{a_n\}$ converges.*

Proof: Suppose $\{a_n\}$ is increasing and bounded above by M. Define $L = \sup_n\{a_n\}$. *How do we know the* sup *exists?* Since L is the least upper bound, $L \leq M$. We will show that $a_n \to L$. Let $\epsilon > 0$ be given. By the definition of sup, there is an $N \in \mathbb{N}$ such that $a_N > L - \epsilon$. Then, for any $n > N$, we have

$$L - \epsilon < a_N < a_n < L$$

Therefore, $|a_n - L| < \epsilon$ for all $n > N$, establishing that $a_n \to L$.   ∎

Since the first finite number of terms do not effect convergence, we could state the result above as: if $\{a_n\}$ is a bounded and *eventually increasing* sequence, then $\{a_n\}$ converges. By *eventually increasing,* we mean that there is an $M \in \mathbb{N}$ so that terms beyond the Mth form an increasing sequence.

Another natural question to ask is when will a set have an accumulation point? The answer comes from a very important theorem depending on the completeness of $\mathbb{R}$ and the previous theorem.

Theorem 2.54 (Bolzano-Weierstrass Theorem) *Every bounded, infinite set of real numbers has an accumulation point.*

Proof: Let E be a bounded, infinite set of real numbers. Then $S \subseteq [a_1, b_1]$ for two real numbers a_1 and b_1. Define $c_1 = (b_1 - a_1)/2$, the midpoint of the interval. Since E is infinite, either $[a_1, c_1]$ or $[c_1, b_1]$ contains infinitely many points of E. Set $[a_2, b_2]$ to be the interval containing an infinite number of points of E. Then $a_1 \leq a_2 \leq b_2 \leq b_1$ and $b_2 - a_2 = (b_1 - a_1)/2$. Continue the process to generate a sequence of intervals $[a_k, b_k]$ where

$$a_1 \leq a_2 \leq \cdots \leq a_k \leq \cdots \leq b_k \leq \cdots \leq b_2 \leq b_1$$

and

$$b_k - a_k = \frac{b_1 - a_1}{2^{k-1}}$$

The sequences $\{a_n\}$ and $\{b_n\}$ are bounded and monotone, and hence convergent by the previous theorem. Suppose $a_n \to A$ and $b_n \to B$. Since

$$b_k - a_k = \frac{b_1 - a_1}{2^{k-1}} \to 0$$

then $\lim a_n = \lim b_n$. That is, $A = B$.

The limit A is an accumulation point of E because every neighborhood about A contains both points a_n and b_n, one of which must be distinct from A. *Why?* Thus, the theorem holds.   ∎

Does the Bolzano-Weierstrass theorem hold if we restrict to rational numbers? No! *Explain why not! Also, look up the phrase "'lion in the desert' proof."*

We are now ready for Cauchy's criterion for demonstrating convergence without knowing a limiting value.

Definition 2.16 (Cauchy Sequence) *A sequence* $\{a_n\}$ *is a* Cauchy sequence *if and only if for any* $\epsilon > 0$ *there is an* $N \in \mathbb{N}$ *such that* $|a_n - a_m| < \epsilon$ *whenever* $n, m > N$.

If we can show a sequence of real numbers to be Cauchy, then we know it converges. We can prove a sequence is Cauchy without knowing its limit!

Theorem 2.55 *A sequence of real numbers is Cauchy if and only if it converges.*

Proof: Suppose $\{a_n\}$ converges; then by a simple application of the triangle inequality, we see that $\{a_n\}$ is Cauchy.

For the other direction, let $\{a_n\}$ be a Cauchy sequence and let $\epsilon > 0$. Set $E = \{a_n | n \in \mathbb{N}\}$. Since $\{a_n\}$ converges, E is bounded. If E is a finite set, then, again using the triangle inequality, $\{a_n\}$ must be eventually constant and so convergent. If E is infinite, then by the Bolzano-Weierstrass theorem, E has an accumulation point, A. Thus every neighborhood of A contains points of the sequence. In particular, the interval $(A - \epsilon/2, A + \epsilon/2)$ contains infinitely many points of the sequence. Since $\{a_n\}$ is Cauchy, there is an N such that for $n, m > N$ we know $|a_n - a_m| < \epsilon/2$. Choose a point $x_m \in (A - \epsilon/2, A + \epsilon/2)$ with $m > N$. *Why must there be one?* Then, for all $n > N$, we have

$$|a_n - A| \leq |a_n - a_m| + |a_m - A| < \frac{\epsilon}{2} + \frac{\epsilon}{2} = \epsilon$$

Thus, $a_n \to A$. ∎

Does the argument above hold if we restrict our universe to rational numbers? No! *Give an example!*

It's not sufficient for consecutive terms to be close for a sequence to be Cauchy; it must be that all terms beyond a certain point are close to each other. For instance, take $h_n = \ln(n)$. Then $h_{n+1} - h_n = \ln(1 + 1/n)$, which goes to zero as $n \to \infty$. *Verify this!* Hence, for any $\epsilon > 0$, there is an N so that $|h_{n+1} - h_n| < \epsilon$, but h_n clearly diverges to infinity and cannot be Cauchy.

The Bolzano-Weierstrass and Cauchy sequence convergence theorems are statements about the completeness of the metric space of real numbers with Euclidean distance. These two theorems show there are no holes in the real line. The metric space of rational numbers with Euclidean distance does not have this property. The rational sequence[3]

$$\{p_n\} = \left\{ 3, \frac{22}{7}, \frac{333}{106}, \frac{355}{113}, \frac{103993}{33102}, \frac{104348}{33215}, \cdots \right\}$$

quickly approaches π. *Check this!* Even though $\{p_n\}$ is Cauchy, it does not converge in $\mathbb{Q}$ but does converge in $\mathbb{R}$.

Another way Cauchy used to describe sequences uses upper and lower limits. Suppose a sequence $\{a_n\}$ is not eventually constant and let $S = \{a_n | n \in \mathbb{N}\}$ be the

---

[3]Maple's `convert(Pi, confrac, 10, 'p')` gives the first 10 terms of $\{p_n\}$.

set of sequence values. If $\{a_n\}$ has a limit L, then L is an accumulation point of S. However, if L is an accumulation point of S, then L may or may not be a limit of the sequence. For example, consider the sequence $c_n = \cos(n\pi) + 1/n$. For this sequence, S has two accumulation points $+1$ and -1 but no limit. However, any accumulation point of S will be the limit of a subsequence of $\{a_n\}$.

The upper limit or *limit superior* will be the largest accumulation point of S, and the lower limit or *limit inferior* will be the smallest.

Definition 2.17 *Let $\{a_n\}$ be a sequence. Then the* limit superior *of $\{a_n\}$ is*

$$\limsup_{n\to\infty} a_n = \inf_{n\to\infty} \sup_{k\geq n} a_k$$

and the limit inferior *of $\{a_n\}$ is*

$$\liminf_{n\to\infty} a_n = \sup_{n\to\infty} \inf_{k\geq n} a_k$$

An alternate definition is based on subsequential limits of $\{a_n\}$. The limit supremum is the supremum of the set of all subsequential limits.

If the largest and smallest accumulation points are the same point, the sequence converges. If not, the sequence diverges. We state this result, leaving the proof to the reader.

Theorem 2.56 *A sequence converges if and only if its limit superior and limit inferior are finite and equal.*

Since $\liminf a_n \leq \limsup a_n$ for all sequences, to show a particular sequence converges we only need to demonstrate $\liminf a_n \geq \limsup a_n$.

Series of Constants

A series is a sequence constructed from another sequence. Bonar and Khoury begin with a clever "pseudo-definition" in their text *Real Infinite Series*: an "infinite series is obtained by taking the terms of an infinite sequence and connecting them with plus signs rather than commas" (Bonar & Khoury, 2006, p. 8).

Definition 2.18 *For any sequence $\{a_k\}$, form the sequence of* partial sums

$$S_n = \sum_{k=p}^{n} a_k$$

for a fixed $p \in \mathbb{Z}$ to generate the series *$\{S_n\}$.*

The infinite series *$S = \sum_{k=p}^{\infty} a_k$ converges if and only if the sequence of partial sums $\{S_n\}$ converges. Otherwise, the series diverges.*

Typically p is 0 or 1 but can be any valid index from the sequence $\{a_n\}$. For the moment, we take $p = 1$. Saying the series converges is equivalent to the limit statement

$$S = \lim_{n\to\infty} S_n \quad \text{or} \quad \sum_{k=1}^{\infty} a_k = \lim_{n\to\infty} \sum_{k=1}^{n} a_k$$

Prior to Cauchy's *Cours d'Analyse*, convergence was not questioned.[4]

■ **EXAMPLE 2.17**

1. Zeno's series

$$\sum_{k=1}^{\infty} \frac{1}{2^n} = 1$$

Since

$$S_n = \frac{1}{2} + \frac{1}{2^2} + \cdots + \frac{1}{2^n}$$

then

$$\frac{1}{2}S_n = \frac{1}{2^2} + \frac{1}{2^3} + \cdots + \frac{1}{2^{n+1}}$$

Then $S_n - (1/2)S_n = (1/2)S_n = 1/2 - 1/2^{n+1}$. Hence, $S_n \to 1$ as $n \to \infty$.

2. The series

$$\sum_{k=1}^{\infty} (-1)^{k+1}$$

diverges.

The sequence of partial sums $\{S_n\} = \{1, 0, 1, 0, 1, 0, \dots\}$ diverges. [Curiously, Leibniz argued this series summed to $1/2$. See Boyer (1959, p. 246).]

3. The harmonic series

$$\sum_{k=1}^{\infty} \frac{1}{k}$$

diverges to infinity.

There have been a number of proofs of the divergence of the harmonic series going back to Oresme (1350), Mengoli (1650), and the Bernoulli's (1689). We'll use Jakob Bernoulli's derivation. [See Struik (1986, p. 320–323) and Dunham (1999, p. 30).]

Let $c \in \mathbb{N}$ be greater than 1. Replacing each term below except the first with the smaller fraction $1/c^2$ gives the inequality

$$\frac{1}{c} + \frac{1}{c+1} + \frac{1}{c+2} + \frac{1}{c+3} + \cdots + \frac{1}{c^2} \geq \frac{1}{c} + (c^2 - c)\frac{1}{c^2} = 1$$

---

[4]In 1703, G. Grandi, at University of Pisa, used the series $1/(1+x) = 1 - x + x^2 - x^3 \pm \cdots$ with $x = 1$ to "show" that $1/2 = 0$ by grouping pairs on the right, writing "this proved mathematically that the world could be created out of nothing," according to Burton (2007, p. 607).

Hence,

$$\sum_{k=1}^{\infty} \frac{1}{k} = 1 + \left(\frac{1}{2} + \frac{1}{3} + \frac{1}{4}\right) + \left(\frac{1}{5} + \frac{1}{6} + \cdots + \frac{1}{25}\right)$$
$$+ \left(\frac{1}{26} + \frac{1}{27} + \cdots + \frac{1}{676}\right) + \cdots$$
$$\geq 1 + 1 + 1 + 1 + \cdots$$

The partial sums increase without bound; the harmonic series diverges.

∎

Geometric series are among the most important. One reason is that we know a simple formula for the sum.

Theorem 2.57 *The geometric series $\sum_{k=0}^{\infty} a\, r^k$ converges to $a/(1-r)$ for $|r| < 1$ and diverges for $|r| \geq 1$.*

Proof: If $|r| \geq 1$; then the general term does not go to zero, so the series must diverge. Suppose that $|r| < 1$. Then $S_n - rS_n = a - a\, r^n$. Solve for S to have

$$S_n = a\, \frac{1 - r^n}{1 - r}$$

Now, since $|r| < 1$, we see that $r^n \to 0$ as $n \to \infty$. Thus

$$\lim_{n \to \infty} S_n = \frac{a}{1 - r} = \sum_{k=0}^{\infty} a\, r^k$$

∎

Let's rephrase Theorem 2.55, the Cauchy criterion for sequences, in terms of series. Cauchy's original statement was in terms of "less than assignable quantities"; Weierstrass popularized the "ϵ-δ" form that we use. [See Burton (2007, p. 608).] Since this theorem is a restatement, we leave the proof to the exercises.

Theorem 2.58 (Cauchy Criterion for Series) *An infinite series $\sum a_n$ converges if and only if for every $\epsilon > 0$ there is an $N \in \mathbb{N}$ such that whenever $n \geq m > N$ we have*

$$|S_n - S_{m-1}| = |a_m + a_{m+1} + \cdots + a_n| = \left|\sum_{k=m}^{n} a_k\right| < \epsilon$$

An immediate corollary of Cauchy's criterion gives another type of convergence. If the series of absolute values of the original terms converges, then we say the series *converges absolutely*. If $\sum a_n$ converges but $\sum |a_n|$ diverges, we say $\sum a_n$ *converges conditionally*.

Corollary 2.59 *If the series $\sum |a_n|$ converges, then $\sum a_n$ also converges.*

The converse of the corollary is not true: the *alternating harmonic series*

$$\sum_{k=1}^{\infty} \frac{(-1)^n}{n} = \ln\left(\frac{1}{2}\right)$$

converges but the harmonic series $\sum 1/n$ diverges to infinity.

If a series converges, the terms must approach zero. *Prove this!* The harmonic series shows the converse is not true. That the terms go to zero does not guarantee convergence. How can we determine if a series converges?

Convergence Tests

A number of tests have been developed to determine series convergence. The advantage of these tests is that we only need to have a formula for the general term of the series. The disadvantage is the tests will not tell us what the sum is for a convergent series.

In order to simplify the statements of the tests, we assume our series have nonnegative terms for the rest of this section.

Comparison Test   The easiest convergence test is to compare the series in question to one we know.

Theorem 2.60 (Comparison Test) *Let $\sum a_n$ and $\sum b_n$ be series. If $0 < a_n \leq b_n$ eventually, that is, when $n > N$, for some $N \in \mathbb{N}$, then*

1. *if $\sum b_n$ converges, then $\sum a_n$ converges.*

2. *if $\sum a_n$ diverges, then $\sum b_n$ diverges.*

Proof: 1. Since the initial segment, the first finite number of terms, does not affect convergence, without loss of generality we can take $n = 1$. Since $a_n \leq b_n$ for each n, the partial sums are also ordered as

$$\sum_{k=1}^{n} a_k \leq \sum_{k=1}^{n} b_k$$

If $\sum b_n$ converges to B, then the partial sums $\sum_{k=1}^{n} a_k \leq B$. These partial sums form a bounded, increasing sequence, and hence converge.

2. On the other hand, if $\sum a_n$ does not converge, then $\sum a_n = \infty$. *Why?* Therefore, the partial sums $\sum_{k=1}^{n} b_k$ must increase without bound. *Why?* Hence, $\sum b_n$ diverges.   ∎

■ **EXAMPLE 2.18**

Determine whether

$$\sum_{k=1}^{\infty} \frac{1}{n^2}$$

converges.

Observe that $1/n^2 \le 2/(n^2 + n)$ for all $n \ge 1$. Now

$$\sum_{k=1}^{n} \frac{2}{n^2 + n} = \sum_{k=1}^{n} \left(\frac{2}{k} - \frac{2}{k+1} \right) = 2 - \frac{2}{n+1} \to 2$$

By the comparison test, $\sum 1/n^2$ must converge. (And converge to something less than 2. *Why?*)   ■

It can be quite difficult to prove $a_n \le b_n$ eventually. A more general approach can be useful.

Limit Comparison Test   If $0 < a_n \le b_n$, then $a_n/b_n \le 1$. Is this relation adequate in the limit, instead of eventually? Yes!

Theorem 2.61 (Limit Comparison Test) *Suppose that $\sum a_n$ and $\sum b_n$ are two series with $a_n \ge 0$ and $b_n > 0$. If $\lim_{n \to \infty} a_n/b_n = L$ where $0 < L < \infty$, then $\sum a_n$ and $\sum b_n$ converge or diverge together.*

Proof: Let $\lim_{n \to \infty} a_n/b_n = L$ and let $\epsilon = L/2$. Then there is an $N \in \mathbb{N}$ so that for all $n > N$

$$\left| \frac{a_n}{b_n} - L \right| < \frac{L}{2}$$

A little algebra allows us to write

$$\frac{L}{2} b_n < a_n < \frac{3L}{2} b_n$$

Applying the comparison test to this expression yields the result.   ■

The cases where $L = 0$ and $L = \infty$ are left to the exercises.

■ **EXAMPLE 2.19**

Determine whether or not the series

$$\sum_{n=1}^{\infty} \frac{n^3 + n + 5}{12n^6 + 2n^4 + 3n + 17}$$

converges.

Since the degree of the numerator is 3 and the denominator is 6, we would choose a series like $\sum 1/n^3$ for comparison. Looking at the limit, we see

$$\lim_{n\to\infty} \frac{n^3 + n + 5}{12n^6 + 2n^4 + 3n + 17} \div \frac{1}{n^3} = \frac{1}{12}$$

Verify this! Since $\sum 1/n^3$ converges, then so does the series in question. ∎

The comparison tests are relatively easy to apply once a suitable series has been chosen. However, there are other tests that handle a series directly.

Integral Test An infinite series can be thought of as an approximation to an improper integral. In Figure 2.17, we see each rectangle has width 1 and height $f(k)$ and is above the curve. Therefore

$$\sum_{k=1}^{\infty} f(k) \geq \int_{1}^{\infty} f(x)\,dx$$

In Figure 2.18, we see each rectangle has width 1 and height $f(k)$ but is now below the curve. Therefore

$$\sum_{k=1}^{\infty} f(k) \leq \int_{0}^{\infty} f(x)\,dx$$

Combining these two observations says the series and integral converge or diverge together.

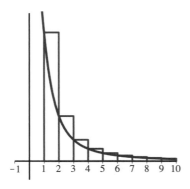

Figure 2.17 Upper Rectangles

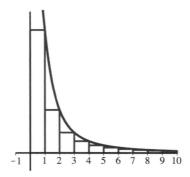

Figure 2.18 Lower Rectangles

Theorem 2.62 (Integral Test) *Let $a : [1, \infty) \to \mathbb{R}$ be continuous, positive, and decreasing with $a_n = a(n)$. Then the series $\sum_{n=1}^{\infty} a_n$ converges if and only if the improper integral $\int_{1}^{\infty} a(x)\,dx$ converges.*

Proof: Since a is continuous on each interval $[k, k+1]$, a is integrable there. Further, since a is decreasing on $[k, k+1]$, then $a_{k+1} \leq a(x) \leq a_k$. Integrating from k to $k+1$ yields

$$a_{k+1} \leq \int_k^{k+1} a(x)\, dx \leq a_k$$

Sum for $k = 1, \ldots, n-1$ to see

$$a_2 + a_3 + \cdots + a_n \leq \int_1^n a(x)\, dx \leq a_1 + a_2 + \cdots + a_{n-1}$$

or $S_n - a_1 \leq \int_1^n a(x)\, dx \leq S_{n-1}$.

If the series converges, then the partial sums S_n are bounded. Hence, the sequence $b_n = \int_1^n a(x)\, dx$ is bounded above and increasing, and thus convergent.

On the other hand, if the integral converges, then the sequence of partial sums S_n is bounded above and increasing and so converges. ∎

■ **EXAMPLE 2.20**

Determine whether or not the harmonic series $\sum_{n=1}^{\infty} 1/n$ converges.

We know full well the harmonic series diverges, but the integral test will give us additional insight. Consider the following computation.

$$\sum_{k=2}^{n} \frac{1}{k} \leq \int_1^n \frac{1}{k}\, dk \leq \sum_{k=1}^{n-1} \frac{1}{k}$$

$$-1 + \sum_{k=1}^{n} \frac{1}{k} \leq \ln(n) \leq \sum_{k=1}^{n-1} \frac{1}{k}$$

Since $\ln(n)$ goes to infinity, not only do we have that the harmonic series diverges, but we also see $\sum_{k=1}^{n} 1/k \approx \ln(n)$. ∎

Unfortunately, an antiderivative can be very difficult or even impossible to find, so we need to keep adding tests to our toolbox. The next test we look at is a favorite for students—it's easy to apply and useful for analyzing power series.

Ratio Test We develop a test that is based on the properties of geometric series.

Theorem 2.63 (Ratio Test) *Let* $\sum a_n$ *be an infinite series of positive terms and set, if the limit exists,*

$$r = \lim_{n \to \infty} \frac{a_{n+1}}{a_n}$$

1. *If* $r < 1$, *then* $\sum a_n$ *converges.*

2. *If* $r = 1$, *the test fails.*

3. *If* $r > 1$, *then* $\sum a_n$ *diverges.*

Proof: 1. Suppose that $r < 1$. Choose a value R where $r < R < 1$. Since r is the limit of a_{n+1}/a_n, then $a_{n+1}/a_n < R$ eventually, i.e., for all large enough n. Thus, $a_{n+1} < Ra_n$. Iterate this relation.

$$a_{n+1} < Ra_n$$
$$a_{n+2} < Ra_{n+1} < R^2 a_n$$
$$a_{n+3} < Ra_{n+2} < R^3 a_n$$
$$\vdots$$
$$a_{n+k} < R^k a_n$$
$$\vdots$$

Summing these inequalities leads to

$$a_{n+1} + a_{n+2} + a_{n+3} + \cdots < a_n(R + R^2 + R^3 + \cdots) = \frac{a_n R}{1 - R}$$

Whence $\sum a_n$ converges. *Why?*

2. Suppose that $r = 1$. Both $\sum 1/n$ and $\sum 1/n^2$ lead to $r = 1$. That $\sum 1/n$ diverges and $\sum 1/n^2$ converges establishes the test is inconclusive.

3. Suppose that $r > 1$. Choose R such that $r > R > 1$. In a similar fashion to the case of $r < 1$, we have the inequality

$$a_{n+1} + a_{n+2} + a_{n+3} + \cdots > \frac{a_n R}{1 - R}$$

for large enough n. Now, however, the geometric series $\sum R^n$ diverges as $R > 1$. The comparison test yields the divergence of $\sum a_n$.   ∎

■ EXAMPLE 2.21

1. Determine whether or not the series $\sum_{n=1}^{\infty} n^2 e^{-n^2}$ converges.

   Simplify the ratio a_{n+1}/a_n.

   $$\frac{a_{n+1}}{a_n} = \frac{(n+1)^2 e^{-(n+1)^2}}{n^2 e^{-n^2}} = \left(1 + \frac{1}{n}\right)^2 e^{-2n-1}$$

   Take the limit as n goes to infinity.

   $$r = \lim_{n\to\infty} \frac{a_{n+1}}{a_n} = \lim_{n\to\infty} \left(1 + \frac{1}{n}\right)^2 e^{-2n-1}$$
   $$= \lim_{n\to\infty} \left(1 + \frac{1}{n}\right)^2 \cdot \lim_{n\to\infty} e^{-2n-1}$$
   $$= 1 \cdot 0 = 0$$

   Since $r = 0 < 1$, the series converges.

2. Determine whether or not the series

$$\sum_{n=2}^{\infty} \left(1 - \frac{1}{n} \right)^n$$

converges.

As above, simplify the ratio a_{n+1}/a_n :

$$\frac{a_{n+1}}{a_n} = \left(1 - \frac{1}{n+1} \right)^{n+1} \cdot \left(1 - \frac{1}{n} \right)^{-n} = \frac{n^{2n+1}}{(n+1)(n^2-1)^n}$$

Verify! Taking the limit as n goes to infinity, we have $\lim a_{n+1}/a_n = 1$. The test fails. *Does this series converge?*

■

As we've seen, the ratio test can fail to resolve simple forms. A stronger test employs roots, rather than ratios.

Root Test   The root test is yet another application of geometric series.

Theorem 2.64 (Root Test) *Let $\sum a_n$ be an infinite series of positive terms, and set, if the limit exists,*

$$\rho = \lim_{n \to \infty} \sqrt[n]{a_n}.$$

1. *If $\rho < 1$, then $\sum a_n$ converges.*

2. *If $\rho = 1$, the test fails.*

3. *If $\rho > 1$, then $\sum a_n$ diverges.*

Proof: 1. Suppose that $\rho < 1$. Choose a value R where $\rho < R < 1$. Since ρ is the limit of $\sqrt[n]{a_n}$, then $\sqrt[n]{a_n} < R$ eventually, which implies $a_n < R^n$ for large enough n. Using the comparison test with the convergent series $\sum R^n$ yields the convergence of $\sum a_n$.

2. Suppose that $\rho = 1$. As in the proof of the ratio test, both $\sum 1/n$ and $\sum 1/n^2$ lead to $\rho = 1$. That $\sum 1/n$ diverges and $\sum 1/n^2$ converges establishes the test is inconclusive.

3. Suppose that $\rho > 1$. For large enough n, it follows that $\sqrt[n]{a_n} > 1$. So, for large n, we have $a_n > 1$, but this implies that $a_n \not\to 0$. Hence $\sum a_n$ diverges.   ■

■ **EXAMPLE 2.22**

1. Determine whether or not the series

$$\sum_{n=1}^{\infty} \left(\frac{2n+1}{3n+2} \right)^n$$

converges.

Applying the root test yields

$$\rho = \lim_{n \to \infty} \sqrt[n]{\left(\frac{2n+1}{3n+2}\right)^n} = \lim_{n \to \infty} \frac{2n+1}{3n+2} = \frac{2}{3}$$

Since $\rho < 1$, the series converges.

2. Determine whether or not the series

$$\sum_{n=2}^{\infty} \left(1 - \frac{1}{n}\right)^n$$

converges.

Check the limit of the nth root.

$$\rho = \lim_{n \to \infty} \sqrt[n]{\left(1 - \frac{1}{n}\right)^n} = \lim_{n \to \infty} \left(1 - \frac{1}{n}\right) = 1$$

The test fails. *Does this series converge?*

   ■

The last test we mention is for *alternating series*. If a series' terms alternates between positive and negative values and the general term goes to zero, then it converges. A standard example is $\sum_{k=1}^{\infty} (-1)^{k+1}/k = \ln(2)$. See Exercise 2.59.

The best way to build facility with analyzing series is with practice. Take several advanced calculus books and look at the exercises following sections on convergence testing for further work. For other versions of the tests we've seen and for more delicate tests such as Cauchy's condensation test, Gauss's test, Raabe's test, etc., see, for example, Bonar & Khoury (2006, Chapter 2), Kosmala (2004, Section 7.2), or Wrede & Spiegel (2002, p. 267–268).

2.6   POINTWISE AND UNIFORM CONVERGENCE

Newton was a master at using infinite series expansions of functions to obtain exciting, new results. In *The Calculus Gallery*, Dunham (2008, p. 8) tells us Newton wrote in 1671,

> all kinds of complicated terms,. . . may be reduced to the class of simple quantities, i.e., to an infinite series . . . , which will thus be freed from those difficulties that in their original form seem'd almost insuperable.

Euler, too, used series to great profit. However, mathematicians of the time "played loosely" with convergence even though Gregory, in 1667, made "a distinction . . . between convergent and divergent series" (Ball, 1912, p. 313). Euler argued that setting $x = 1$ in $1/(1 + x)^2 = 1 - 2x + 3x^2 - 4x^3 \pm \cdots$ showed that $1 - 2 + 3 - 4 \pm \cdots = 1/4$. [See Boyer (1959, p. 246).] Today, we don't agree. The question of what values and actions are permissible is again one of convergence.

We will see, just as we did for continuity, sequences and series of functions have two types of convergence, *pointwise* and *uniform*. The same heuristic as before is valid: pointwise convergence properties vary with the value of x, while uniform convergence properties are independent of the value of x. Weierstrass, the "father of modern analysis," figures prominently in the development of the concept of uniform convergence.

Pointwise Convergence

We generalize from sequences and series of constants to sequences and series of functions.

Definition 2.19 *Let* $f_n : D \to \mathbb{R}$ *for* $n \in \mathbb{N}$ *be a collection of functions on the common domain* D. *Then* $\{f_n\}$ *forms a* sequence of functions *and* $\sum f_n(x)$ *forms a* series of functions.

For any particular value $x = a \in D$, we form the sequence of constants $\{f_n(a)\}$ and the series of constants $\sum f_n(a)$, as shown in Figures 2.19 and 2.20, that were studied in the previous section.

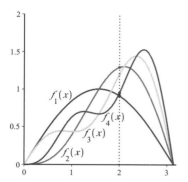

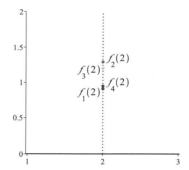

Figure 2.19   Sequence of Functions   **Figure 2.20**   Associated Sequence of Points

Definition 2.20 *A sequence* $\{f_n\}$ *or series* $\sum f_n(x)$ *of functions* converges pointwise *on a set* A *to a function* f *if and only if for each* $a \in A$ *the sequence* $\{f_n(a)\}$ *or series* $\sum f_n(a)$ *of constants converges, respectively.*

We say f is the *pointwise limit* of the sequence or series and write

$$\lim_{n \to \infty} f_n(x) = f(x) \qquad \text{or} \qquad \sum_{k=1}^{\infty} f_n(x) = f(x)$$

Phrasing Definition 2.20 in terms of ϵ and N reads $f_n(x) \to f(x)$ on A if and only if given $a \in A$ and any $\epsilon > 0$ there is an $N = N(x,\epsilon) \in \mathbb{N}$ such that, whenever $n > N$, then $|f_n(a) - f(a)| < \epsilon$. We use the notation $N = N(x,\epsilon)$ to emphasize that the choice of N depends on both ϵ and a. Hence, the name *pointwise convergence*.

■ **EXAMPLE 2.23   Abel (1826)**

Let

$$f_n(x) = (-1)^{n+1} \frac{\sin(nx)}{n}$$

for $x \in [-\pi,\pi]$ and set

$$S_n(x) = \sum_{k=1}^{n} f_n(x)$$

Then $S_n(x)$ converges pointwise to $f(x) = x/2$ on $(-\pi,\pi)$ and $f(\pm\pi) = 0$.

Fix an $x \in [-\pi,\pi]$. The alternating series test (Exercise 2.59) proves convergence. *Do it!* Even though each partial sum S_n is a continuous function (*Why?*), the pointwise limit is not. Figure 2.21 shows S_1, S_4, and S_{15} along with the discontinuous limit function f.   ■

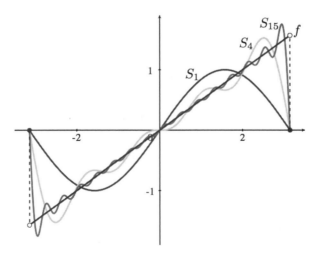

Figure 2.21   Abel's Example of Pointwise Convergence

We learn from Abel's example that the pointwise limit of continuous functions need not be continuous. We also see pointwise limits do not preserve differentiability from the same example.

■ **EXAMPLE 2.24**

Define the sequence of function $\{f_n\}$ on $[0,1]$ by

$$
f_n(x) = \begin{cases}
2n^2x & x \leq \dfrac{1}{2n} \\[2mm]
2n(1-nx) & \dfrac{1}{2n} < x \leq \dfrac{1}{n} \\[2mm]
0 & \dfrac{1}{n} < x
\end{cases}.
$$

The graph in Figure 2.22 shows several f_ns.

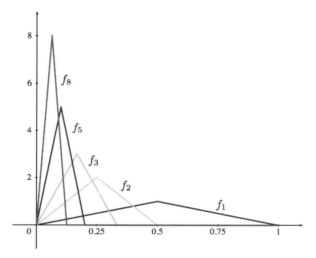

Figure 2.22   Marching Triangles

Since $1/n \to 0$, we see that $f_n(x) \to 0$. *Show this!*
Consider

$$
\int_0^1 \lim_{n\to\infty} f_n(x)\, dx \qquad \text{and} \qquad \lim_{n\to\infty} \int_0^1 f_n(x)\, dx
$$

Since $f_n(x) \to 0$, then

$$
\int_0^1 \lim_{n\to\infty} f_n(x)\, dx = \int_0^1 0\, dx = 0
$$

On the other hand, each f_n creates a triangle with base $1/n$ and height n. Thus, each area is $1/2$. So

$$
\lim_{n\to\infty} \int_0^1 f_n(x)\, dx = \lim_{n\to\infty} \frac{1}{2} = \frac{1}{2}
$$

The limit of the integrals is not equal to the integral of the limit.   ■

Integrability also fails to be preserved by pointwise convergence. The problem is the interchange of infinite processes. Each of continuity, differentiability, and integrability requires passing to a limit. Exchanging the order of $\lim_{n\to\infty}$ and $\lim_{x\to a}$, etc., isn't working. What do we need? Uniformity in x!

Uniform Convergence

Weierstrass solved the problem of interchanging limiting processes in convergence by developing the concept of *uniform convergence* around 1841 (although he didn't publish it until 1894) (Dunham, 2008, p. 136). The essential point is removing the specific value of the domain from the discussion, the property holds equally for all x, or "uniformly." We will separate the definitions for sequences and series for ease of reading.

First, uniform convergence of sequences.

Definition 2.21 (Uniform Convergence of Sequences) *A sequence of functions* $\{f_n\}$ *converges uniformly to a function f on a set A if and only if for any $\epsilon > 0$ there is an $N = N(\epsilon) \in \mathbb{N}$ so that, whenever $n > N$, then $|f_n(x) - f(x)| < \epsilon$ for every $x \in A$.*

Now, uniform convergence of series.

Definition 2.22 (Uniform Convergence of Series) *A series of functions $\sum f_n$ converges uniformly to a function f on a set A if and only if for any $\epsilon > 0$ there is an $N = N(\epsilon) \in \mathbb{N}$ so that, whenever $n > N$, then $\left|\sum_{k=1}^{n} f_k(x) - f(x)\right| < \epsilon$ for every $x \in A$.*

In both definitions, we have written $N = N(\epsilon)$ to emphasize the independence of N from any value of $x \in A$.

Figure 2.23 shows the sequence $f_n(x) = x^n$ and a shaded ϵ-strip about the limit function $f(x) = 0$ for $x \neq 1$ and $f(1) = 1$ on the set $A = [0, 1]$. [Note the portion of the ϵ-strip at the point $(1, 1)$.] Since f_n leaves the ϵ-strip near 1 for all n, this image does not show uniform convergence. *Does f_n converge pointwise to f?* Figure 2.24 shows the sequence $g_n(x) = \sin^n(x)/\ln(n + 1)$ converging uniformly to $g(x) = 0$ on $[0, \pi]$. Soon, all the g_n's, for n large enough, will be captured by the shaded ϵ-strip. We must be able to enclose all f_n in the ϵ-strip about f for all $n > N$ to have uniform convergence. *Does Example 2.23, Abel's example, show uniform or pointwise convergence?*

A simple test for uniform convergence follows immediately from the definition; the proof is left to the reader.

Theorem 2.65 *Let f_n converge pointwise to f on A and set*

$$M_n = \sup_{x \in A} |f_n(x) - f(x)|.$$

Then f_n converges to f uniformly on A if and only if M_n converges to 0.

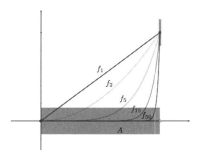

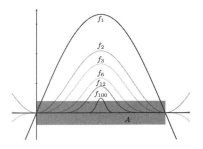

Figure 2.23   Nonuniform Convergence.   **Figure 2.24**   Uniform Convergence.

We have met the geometric series $\sum x^n$ several times and know that it converges for $|x| < 1$. Is this pointwise or uniform convergence? We know there are problems at both endpoints, -1 and 1. If we stay away from the endpoints, then f_n converges uniformly on intervals $[-r, r]$ where $0 < r < 1$. Dini's theorem (see Exercise 2.67) tells us that since $\sum x^n$ is a sequence of monotonic, continuous functions converging pointwise on a closed and bounded set $[0, r]$ then the series converges uniformly. Can this result also be applied to $\sum x^n$ on $[-r, 0]$?

Uniform convergence removes the particular value of x from consideration. This property is just what we need to be able to exchange the infinite processes in continuity.

Theorem 2.66 *If a sequence or series of continuous functions converges uniformly on a set, then the limit function is continuous on that set.*

We do the proof for sequences and leave series to the reader.

Proof: Let $\{f_n\}$ be a sequence of continuous functions converging uniformly to f on A. Let $\epsilon > 0$ and let $a \in A$. Since f_n converges to f uniformly, there is an $N \in \mathbb{N}$ such that for all $n > N$ and all $x \in A$

$$|f_n(x) - f(x)| < \frac{\epsilon}{3}$$

Fix an $m > N$. Since f_m is continuous, there is a $\delta > 0$ such that if $x \in A$ and $|x - a| < \delta$ then

$$|f_m(x) - f_m(a)| < \frac{\epsilon}{3}$$

Now, let $x \in A$ and $|x - a| < \delta$; then

$$|f(x) - f(a)| \le |f(x) - f_m(x)| + |f_m(x) - f_m(a)| + |f_m(a) - f(a)|$$
$$\le \frac{\epsilon}{3} + \frac{\epsilon}{3} + \frac{\epsilon}{3} = \epsilon$$

Hence, f is continuous on A.   ∎

The converse of the theorem is not true. A sequence of very discontinuous functions can converge to a continuous limit. For example, take

$$D_n(x) = \begin{cases} \dfrac{1}{n} & x \in \mathbb{Q} \\ 0 & \text{otherwise} \end{cases}$$

Then D_n converges uniformly to 0, but each D_n is nowhere continuous.

However, the contrapositive of the theorem is quite useful. The continuous functions x^n converge to a discontinuous limit on $[0, 1]$; therefore the convergence cannot be uniform.

Uniform convergence also preserves integrability.

Theorem 2.67 *If a sequence $\{f_n\}$ or series $\sum g_n$ of continuous functions converges uniformly on an interval $[a, b]$, then*

$$\lim_{n \to \infty} \int_a^b f_n(x)\, dx = \int_a^b \left(\lim_{n \to \infty} f_n(x) \right) dx$$

or

$$\sum_{n=1}^{\infty} \int_a^b g_n(x)\, dx = \int_a^b \left(\sum_{n=1}^{\infty} g_n(x) \right) dx$$

Proof: Let $\{f_n\}$ be a sequence of continuous functions converging uniformly to f on $[a, b]$ and let $\epsilon > 0$. Since the functions are all continuous, they are all integrable on $[a, b]$. In particular, each difference $f_n - f$ is integrable on $[a, b]$.

Since $f_n \to f$ uniformly, there is an $N \in \mathbb{N}$ such that for any $x \in [a, b]$ and for all $n > N$ we have

$$|f_n(x) - f(x)| < \frac{\epsilon}{b - a}$$

Let $n > N$; then

$$\left| \int_a^b f_n(x)\, dx - \int_a^b f(x)\, dx \right| = \left| \int_a^b (f_n(x)\, dx - f(x))\, dx \right|$$

$$\leq \int_a^b |f_n(x)\, dx - f(x)|\, dx$$

$$< \int_a^b \frac{\epsilon}{b - a}\, dx = \epsilon$$

Since this inequality holds for all $n > N$, we have

$$\lim_{n \to \infty} \int_a^b f_n(x)\, dx = \int_a^b f(x)\, dx$$

and the result follows. ∎

As before, writing the details of the proof for series is left to the reader.

Since the sequence of integrals of the "marching triangles" in Example 2.24 did not match the integral of the limit, the convergence to zero could not be uniform. Let's consider an example of integrals under uniform convergence.

■ **EXAMPLE 2.25**

Let

$$f_n(x) = \frac{x}{1 + nx^2}$$

on $[0, 1]$.

Graphs of several f_n are shown in Figure 2.25.

A little calculus shows that $f_n(x)$ has a maximum of $1/(2\sqrt{n})$ occurring at $1/\sqrt{n}$. *Do it!* Then, for a given $\epsilon > 0$, choosing $N \geq 1/(4\epsilon^2)$ shows uniform convergence to $f(x) = 0$. *Why?* For each n, the integral

$$\int_0^1 f_n(x)\, dx = \int_0^1 \frac{x}{1 + nx^2}\, dx = \frac{1}{2} \frac{\ln(n + 1)}{n}$$

which goes to 0 as n goes to infinity as required by the theorem.   ■

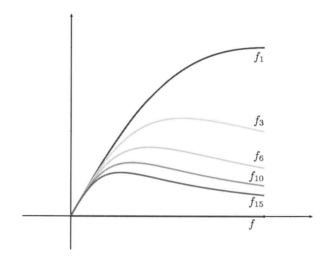

Figure 2.25 The Sequence $f_n(x) = x/(1 + nx^2)$.

What about differentiation and uniform convergence? Does this property follow as simply as continuity and integrability?

Consider the sequence above $f_n(x) = x/(1 + nx^2)$ at $x = 0$. For each n, we have

$$f_n'(x) = \frac{1 - nx^2}{(1 + nx^2)^2}$$

so $f'_n(0) = 1$. *Verify this!* But $f'(0) = 0$. By itself, uniform convergence is not enough to guarantee differentiation.

Theorem 2.68 *Let $\{f_n\}$ be a sequence of functions defined on the interval $[a, b]$ such that*

1. *each f_n is continuously differentiable on $[a, b]$,*

2. *f_n converges pointwise to f on $[a, b]$, and*

3. *the sequence $\{f'_n\}$ converges uniformly on $[a, b]$.*

Then

1. *f is continuously differentiable on $[a, b]$ and*

2. *f'_n converges uniformly to f' on $[a, b]$.*

Proof: Since each f'_n is continuous, the fundamental theorem tells us

$$\int_a^x f'_n(t)\, dt = f_n(x) - f_n(a)$$

for any $x \in [a, b]$. Since f'_n converges uniformly to, say, g, we see

$$\lim_{n \to \infty} \int_a^x f'_n(t)\, dt = \int_a^x \lim_{n \to \infty} f'_n(t)\, dt$$

that is,

$$\lim_{n \to \infty} (f_n(x) - f_n(a)) = \int_a^x \lim_{n \to \infty} f'_n(t)\, dt$$

which, because f_n converges to f, becomes

$$f(x) - f(a) = \int_a^x \lim_{n \to \infty} f'_n(t)\, dt$$

Differentiate both sides to yield

$$f'(x) = \frac{d}{dx} \int_a^x \lim_{n \to \infty} f'_n(t)\, dt = \lim_{n \to \infty} f'_n(x)$$

Thus f'_n converges uniformly to f'. ∎

To see versions of the differentiability theorem with weaker hypotheses, see Kosmala (2004, p. 349) or Rudin (1976, p. 152).

Our next theorem concerning uniform convergence is the extraordinarily useful convergence test due to Weierstrass.

Theorem 2.69 (Weierstrass M-Test) *Let $\{f_n\}$ be a sequence of functions and $\{M_n\}$ be a sequence of constants such that $|f_n(x)| \le M_n$ for all $x \in D$ and for each $n \in \mathbb{N}$. If $\sum M_n$ converges, then $\sum f_n$ converges uniformly and absolutely on D.*

The proof follows easily from the Cauchy criterion for convergence and is left to the exercises.

■ **EXAMPLE 2.26**

Test the series $\sum_{k=1}^{\infty} \cos(nx)/n^2$ for convergence on $\mathbb{R}$.
The cosine function is bounded by 1 on $\mathbb{R}$; therefore

$$\left| \frac{\cos(nx)}{n^2} \right| \leq \frac{1}{n^2}$$

Since $\sum 1/n^2$ converges, then $\sum \cos(nx)/n^2$ converges uniformly and absolutely on $\mathbb{R}$. *What powers other than n^2 would work here?*   ■

To end our investigations into convergence, we state a last theorem on uniform convergence without proof. Weierstrass's approximation theorem shows, in one sense, how polynomials dominate.

Theorem 2.70 (Weierstrass Approximation Theorem) *If the function $f : [a, b] \to \mathbb{R}$ is continuous, there is a sequence of polynomials p_n converging to f uniformly on $[a, b]$.*

Power and Taylor Series

The history of power and Taylor series is very long. While mathematicians from Archimedes' time forward used series of constants, these were limited to specific cases. Oresme (1350), Suiseth (1350), Viète (1593), and Mengoli (1650) used the series expansion $1/(1 - r) = 1 + r + r^2 + \cdots$, among others, but did not understand the expressions as functions.[5] It was in the 1600s that Newton, Gregory, and Mercator focused on series representations of algebraic expressions. However, none of them, as far as we know, attempted to represent an arbitrary function. Newton developed the general binomial expansion of $(1 + x)^p$ for rational p in 1666 (Hairer & Wanner, 1996, p. 24), although he did not publish the result until much later. Gregory sent a letter to Collins on February 16, 1671, that included power series expansions for tangent, arctangent, secant, arcsecant, and several logarithm compositions (Dehn & Hellinger, 1943, p. 149). Mercator published a power series for $\ln(1 + x)$ in 1668 (Stillwell, 1989, p. 120). These results became the foundation for expanding the role of power series. Taylor's 1715 text *Methodus incrementorum directa et inversa* introduced Taylor series. [Although he had written a letter to Machin on July 26, 1712, explaining the concept and his motivation according to O'Connor & Robertson (2008).] However, it wasn't until Lagrange's (1797) text *Théorie Des Fonctions Analytiques* attempting to base calculus on power series to avoid the infinitesimal problem that the significance of Taylor series was widely understood. [See Boyer (1959, p. 260).] Read Kahane's (2000) intriguing article "A Century of Interplay Between Taylor Series, Fourier Series and Brownian Motion" for a description of work on Taylor series over various time periods.

---

[5]Madhava developed a theory of expansions of functions in infinite series in India in the fourteenth century; his work was not known in the West.

Let's start with power series.

Definition 2.23 (Power Series) *A* power series *is a function of the form*

$$f(x) = \sum_{k=0}^{\infty} a_n x^n$$

or, centered at $x = a$,

$$f(x) = \sum_{k=0}^{\infty} a_n (x - a)^n$$

If a power series converges for some specific value of x, say x_0, then it must converge for any value x such that $|x - a| < |x_0 - a|$. *Why?* We call the largest R, where $0 \le R \le \infty$, such that f converges for all $x \in (a - R, a + R)$ the *radius of convergence*. If R is the radius of convergence, then $(a - R, a + R)$ is the *interval of convergence*. The integral of convergence may or may not include its endpoints. Inside the interval, we have very nice behavior. For the rest of our discussion of power series, we'll let $a = 0$ without loss of generality.

Theorem 2.71 *Suppose R is the radius of convergence of the power series*

$$f(x) = \sum_{n=0}^{\infty} a_n x^n$$

where $0 \le R \le \infty$. Then the power series $f(x)$

1. *converges absolutely for $|x| < R$.*

2. *converges uniformly for $|x| \le r$ where $0 < r < R$.*

3. *diverges for $|x| > R$.*

Proof: Let $0 < r < R$ be given and let $x \in [-r, r]$. Then

$$|a_n x^n| \le |a_n r^n|$$

Using the root test, we have that $\sum a_n r^n$ converges. Therefore, by the Weierstrass M-test, $\sum a_n x^n$ converges uniformly.

For $|x| > R$, we see that f diverges by the definition of R.   ∎

It's often possible to find the radius of convergence of a power series using the ratio or root tests. A computer algebra system makes the task much easier.

■ **EXAMPLE 2.27**

Find the radius of convergence for the power series

$$\sum_{n=1}^{\infty} \frac{1}{k} \left(\frac{x}{2}\right)^k$$

Form the ratio $|a_{n+1}/a_n|$ and simplify.

$$\left|\frac{a_{n+1}}{a_n}\right| = \frac{1}{2} \frac{k}{k+1} |x|$$

Now let n go to infinity.

$$\lim_{n \to \infty} \left|\frac{a_{n+1}}{a_n}\right| = \frac{1}{2} |x|$$

The ratio test tells us this series converges for $|x|/2 < 1$ or $|x| < 2$. So the radius of convergence is 2. *Check the endpoints $x = \pm 2$ for convergence to determine the interval of convergence!*   ■

We know that series behave well with respect to continuity, differentiation (mostly), and integration under uniform convergence. Power series work very well.

Theorem 2.72 *Suppose the power series*

$$f(x) = \sum_{n=0}^{\infty} a_n x^n$$

converges on $(-R, R)$ for some radius of convergence $R > 0$. Then

1. *f is continuous on $(-R, R)$.*

2. *f is differentiable on $(-R, R)$ and*

$$f'(x) = \sum_{n=1}^{\infty} n a_n x^{n-1}$$

3. *f is integrable on $(-R, R)$ and*

$$\int_0^x f(t)\, dt = \sum_{n=0}^{\infty} \frac{a_n}{n+1} x^{n+1}$$

Proof: We begin with 2. Since $\sqrt[n]{n} \to 1$ as $n \to \infty$, then

$$\lim_{n \to \infty} \sqrt[n]{n|a_n|} = \lim_{n \to \infty} \sqrt[n]{|a_n|}$$

Hence, the root test implies the series of derivatives has the same interval of convergence. Apply the previous theorem to have uniform convergence and then 2. follows on $[-r, r]$. Since $r < R$ is arbitrary, then 2 holds on $(-R, R)$.

Part 1 follows immediately from 2. *Why?*

Part 3 is shown by an analogous argument. *How?*   ∎

We can be more specific with a power series' derivatives. Iterate the result above to prove the following corollary.

Corollary 2.73 *The power series* $f(x) = \sum_{n=0}^{\infty} a_n x^n$ *has derivatives of all orders in its interval of convergence, and*

$$f^{(k)}(x) = \sum_{n=k}^{\infty} n(n-1) \cdots (n-k+1) a_n x^{n-k}$$

Further,

$$f^{(k)}(0) = k! a_k$$

for $k = 0, 1, 2, \ldots$.

■ EXAMPLE 2.28

We know that $f(x) = \sum_{n=0}^{\infty} x^n$ converges to $1/(1-x)$ for $|x| < 1$. In light of Theorem 2.72, we see

1. $f'(x) = \sum_{n=1}^{\infty} n x^{n-1}$. Hence, $f'(x) = \dfrac{1}{(1-x)^2} = \sum_{n=1}^{\infty} n x^{n-1}$ for $|x| < 1$.

2. $\displaystyle\int_0^x f(t)\, dt = \sum_{n=0}^{\infty} \frac{x^{n+1}}{n+1}$. Hence, $\displaystyle\int_0^x f(t)\, dt = -\ln(1-x) = \sum_{k=1}^{\infty} \frac{x^k}{k}$ for $|x| < 1$.   ∎

Before we turn to Taylor series, we need to know if it is possible for two different series to represent the same function on an interval.

Theorem 2.74 *Series representation is unique on the interval of convergence.*

Proof: Suppose f is represented by two power series on the interval of convergence $(-R, R)$. Let $\sum a_n x^n$ and $\sum b_n x^n$ be two power series representations of f. Then

$$f(0) = \sum_{n=0}^{\infty} a_n 0^n = \sum_{n=0}^{\infty} b_n 0^n \implies a_0 = b_0$$

and

$$f'(0) = \sum_{n=1}^{\infty} n a_n 0^{n-1} = \sum_{n=1}^{\infty} n b_n 0^{n-1} \implies a_1 = b_1$$

and so forth. We see that $a_n = b_n$ for all n. Thus, the two series are identical.   ∎

In Section 2.9, we looked at Taylor polynomials as an application of differentiation. Now we see Taylor series as special power series. The motivation is Corollary 2.73 stating $a_k = f^{(k)}(0)/k!$ for $k = 0, 1, 2, \ldots$. Combining this relation on coefficients with the uniqueness of series representation proves Taylor's theorem. We'll state the form centered at $x = a$ noting that a Taylor series centered at $x = 0$ is called a Maclaurin series.

Theorem 2.75 (Taylor Series) *Suppose that $f(x) = \sum_{n=0}^{\infty} a_n(x-a)^n$ has a nonzero radius of convergence R. Then f is infinitely differentiable on $(a - R, a + R)$, and the coefficients a_n are given by*

$$a_n = \frac{f^{(n)}(a)}{n}$$

for $n = 0, 1, 2, \ldots$.

Look back at Example 1.13 for a set of important Taylor and Maclaurin series.

A function whose Taylor series converges to it on a neighborhood of a point a is called *analytic at a*. Our final theorem of this section is Bernstein's surprising result on Taylor series of analytic functions.

Theorem 2.76 (Bernstein's Theorem) *If f and all its derivatives of all orders are nonnegative in an interval (a, b), then f is analytic on (a, b). That is, f has a convergent Taylor series representation on (a, b).*

The proof is based on showing the Taylor remainders for f are bounded by $r^n \cdot f(b)$, where $r = (x - a)^n/(b - a)^n$ is less than one. The remainders then go to zero as n goes to infinity.

The ability to differentiate a function in general led to Taylor series. To go further from here, replacing the powers x^n with trigonometric terms $\cos(nx)$ and $\sin(nx)$ yields Fourier series. Fourier series led to advances in integration by Riemann and Lebesgue. In the next chapter, we'll study Lebesgue's methods.

Summary

We have now put strong foundations under the elementary calculus topics of Chapter 1. Studying real analysis while paying attention to its history allows us to see not only the connections between seemingly disparate topics, but also the development over time and to understand the causes that motivated mathematicians' work. We've seen the tangent problem led to derivatives, which in turn led to foundational definitions. We've seen quadrature led to integration. We've noted the epiphanies of Newton and Leibniz recognizing differentiation and integration as inverse operations and seen how useful the concepts were. Even though we've re-engineered the calculus, carefully defining concepts and proving results, we've still just begun. The next steps will take us into the twentieth century to another mathematical revolution initiated by Henri Lebesgue. The adventure continues.

EXERCISES

2.1 Prove Theorem 2.1, the cancellation Laws.

2.2 Prove Theorem 2.5. Every open interval (a, b) contains both a rational and an irrational number.

2.3 Define $d : \mathbb{R} \to \mathbb{R}$ by

$$d(x, y) = \frac{|x - y|}{1 + |x - y|}$$

a) Show that d is a metric on $\mathbb{R}$.
b) Show d is bounded below by 0 and above by 1.
c) Describe the set

$$N_{1/2}(0) = \{x \mid d(x, 0) < 1/2\}$$

2.4 Carefully prove:
a) An open interval is an open set.
b) A closed interval is a closed set.
c) The half-open interval $[0, 1)$ is neither open nor closed.

2.5 Prove or disprove:
a) Every point of an open set is an accumulation point.
b) Every point of a closed interval is an accumulation point.

2.6 Show that every point of a closed set is either an accumulation point or an isolated point.

2.7 Prove that a point x is an isolated point of a set S if and only if there is a deleted neighborhood $N_r'(x)$ such that $N_r'(x) \cap S = \emptyset$.

2.8 Prove every open set O on the real line is the union of countably many pairwise disjoint open intervals I_n.
a) Define the relation $\sim$ by $a \sim b$ if the whole line segment between a and b is contained in O. Show that $\sim$ is an equivalence relation. Then O is the union of the pairwise disjoint equivalence classes of $\sim$. Denote the equivalent classes by I_n.
b) Show each I_n is an interval. (Take any two points $x, y \in I_n$. Being in the same equivalence classes, x and y must be related.)
c) Show each I_n is open.
d) Show there are only countably many $I - n$. (Each of the disjoint I_n's must contain at least one different rational number.)

2.9 Show the following statements are equivalent.

1. If $x \in N_\delta'(a) \cap \operatorname{dom}(f)$, then $f(x) \in N_\epsilon(L)$.

2. If $x \in \operatorname{dom}(f)$ and $0 < |x - a| < \delta$, then $|f(x) - L| < \epsilon$.

2.10 Prove Theorem 2.9, the algebra of limits.

2.11 Determine whether

$$\lim_{x \to 0} \sqrt{x^4 - x^2}$$

exists. If it does, find the value. If it doesn't, explain why not.

2.12 Prove Theorem 2.10, the sandwich theorem.

2.13 Suppose f is an odd function that has a limit at $a = 0$. Find $\lim_{x \to 0} f(x)$.

2.14 Let $n \in \mathbb{N}$. Find, if it exists,

$$\lim_{x \to 1} \frac{x^n - 1}{x - 1}$$

2.15 Use a graphing utility to find two functions $L(x)$ and $U(x)$ with $L(0) =$

$U(0)$ and such that

$$L(x) \leq \frac{1 - \cos(x)}{x^2} \leq U(x)$$

on $[0, 1]$. Use L and U to determine

$$\lim_{x \to 0} \frac{1 - \cos(x)}{x^2}$$

2.16   Consider the sum $f + g$.
  a) Suppose f is continuous at a but g is not. Is $f + g$ continuous or not?
  b) Can $f + g$ be continuous at a if both f and g are discontinuous there?

2.17   Prove Theorem 2.13, the composition of continuous functions is continuous.

2.18   Define a so that the function

$$T(\theta) = \begin{cases} \frac{1 - \cos(\theta)}{\theta^2} & \theta \neq 0 \\ a & \theta = 0 \end{cases}$$

is continuous, if possible.

2.19   Suppose f is an odd function that is continuous at $a = 0$. Find $f(0)$, if possible.

2.20   Suppose f is an even function that is continuous at $a = 0$. Find $f(0)$, if possible.

2.21   Discuss the continuity of the function

$$\Phi(x) = \begin{cases} x & x \in \mathbb{Q} \\ 0 & x \notin \mathbb{Q} \end{cases}$$

2.22   Discuss the continuity of the function

$$\Psi(x) = \begin{cases} x^2 & x \in \mathbb{Q} \\ 0 & x \notin \mathbb{Q} \end{cases}$$

2.23   The Heaviside function is defined by

$$U(x) = \begin{cases} 1 & x > 0 \\ 0 & x \leq 0 \end{cases}$$

  a) Show U is not differentiable at zero.
  b) Graph $U'(x)$ for $x \neq 0$.
  c) Determine $\lim_{x \to 0} U'(x)$.

2.24   Set $h(x) = x^p \sin(1/x)$ if $x \neq 0$ and $h(0) = 0$.
  a) Show $h'(0)$ does not exist when $p \leq 1$.
  b) Find $h'(0)$ when $p > 1$.
  c) Prove that h is continuous if $p > 0$.

2.25   Let $k_n(x) = |x| \cdot x^n$.
  a) Discuss the continuity of $k_2'''(0)$.
  b) Discuss the continuity of $k_3'''(0)$.

2.26   Prove Theorem 2.24, the algebra of derivatives.

2.27   Give examples showing the three hypotheses of Theorem 2.28, Rolle's theorem, are necessary.

2.28   Prove Theorem 2.28, Rolle's theorem.

2.29   Use the mean value theorem to prove the statement: If f is continuous on $[a, b]$ and $f'(x) = 0$ for all x, then f is a constant function.

2.30   In Example 2.9, we let $f(x) = \cos(x^2/5)$ and $g(x) = \sin(x/3)$ on $[0, 2\pi]$. Use a computer algebra system to find the three values of $c \in (0, 2\pi)$ satisfying Theorem 2.31, Cauchy's mean value theorem.

2.31 Let

$$f_a(x) = \begin{cases} x^a \sin\left(\dfrac{1}{x}\right) & x \neq 0 \\ 0 & x = 0 \end{cases}$$

a) Find $f_a'(x)$ for $x \neq 0$.

b) Find $f'(0)$ using the definition of derivative for $a > 1$.

c) Find $f'(0)$ using the definition of derivative for $a \leq 1$.

d) Does $\lim_{x \to 0} f'(x) = f'(0)$?

2.32 Discuss the differentiability of the function

$$\Phi(x) = \begin{cases} x & x \in \mathbb{Q} \\ 0 & x \notin \mathbb{Q} \end{cases}$$

2.33 Discuss the differentiability of the function

$$\Psi(x) = \begin{cases} x^2 & x \in \mathbb{Q} \\ 0 & x \notin \mathbb{Q} \end{cases}$$

2.34 Apply l'Hôpitals rule to find

a) $\lim_{x \to 0} \dfrac{\sin(x)}{x}$

b) $\lim_{x \to 0} \dfrac{1 - \cos(x)}{x^2}$

c) $\lim_{x \to 0} \dfrac{\tan(x) - x}{x^3}$

2.35 Find the flaw in the "proof" below showing

$$\lim_{x \to 2} \frac{3x^2 - 5x - 2}{x^2 - 2x} = 3$$

Proof. Apply l'Hôpitals rule to see

$$\lim_{x \to 2} \frac{3x^2 - 5x - 2}{x^2 - 2x} = \lim_{x \to 2} \frac{6x - 5}{2x - 2}$$

Again apply l'Hôpitals rule; then

$$\lim_{x \to 2} \frac{6x - 5}{2x - 2} = \lim_{x \to 2} \frac{6}{2} = 3$$

What is the correct value of the limit?

2.36 The absolute value function has a derivative with a simple jump discontinuity at the origin. Explain why this does not contradict Theorem 2.33, Darboux's intermediate value theorem for derivatives.

2.37 Use induction to prove the Leibniz rule for derivatives. If f and g are n-times differentiable, then

$$(fg)^{(n)} = \sum_{k=0}^{n} \binom{n}{k} f^{(k)} g^{(n-k)}$$

2.38 Use Taylor expansions to find

a) $\lim_{x \to 0} \dfrac{\sin(x)}{x}$

b) $\lim_{x \to 0} \dfrac{1 - \cos(x)}{x^2}$

c) $\lim_{x \to 0} \dfrac{\tan(x) - x}{x^3}$

2.39 Compute the Riemann integral $\int_0^1 x^3\, dx$ using the definition.

2.40 Prove Theorem 2.39, the algebra of Riemann integrals.

2.41 Use a computer algebra system to plot the function $Si(x) = \int_0^x \sin(t)/t\, dt$ for $x \in [0, 1]$. Is Si continuous?

2.42 Prove the *Cauchy-Bunyakovsky-Schwarz inequality for integrals*: If f and g are Riemann integrable on $[a, b]$, then

$$\left[\int_a^b f(x)g(x)\, dx \right]^2$$
$$\leq \left[\int_a^b f(x)\, dx \right]^2 \left[\int_a^b g(x)\, dx \right]^2$$

2.43 If f is nonnegative on $[a, b]$, then prove $\int_a^b f(x)\, dx = 0$ if and only if $f(x) \equiv 0$.

2.44   What choice of α makes the Riemann integral a special case of the Riemann-Stieltjes integral?

2.45   Let α be monotone increasing on $[a, b]$. Prove

$$\int_a^b d\alpha = \alpha(b) - \alpha(a)$$

2.46   Let μ be the unit step function and define f by

$$f(x) = \begin{cases} 1 + x & \text{if } x \in \mathbb{Q} \\ 1 - x & \text{otherwise} \end{cases}$$

In light of Example 2.14, calculate the value of $\int_{-1}^1 f \, d\mu$, if possible.

2.47   Use Example 2.14 to show

$$\sum_{k=1}^n f(k) = \int_0^n f(x) \, d\lfloor x \rfloor$$

2.48   Prove Theorem 2.47, part 2,

$$\left| \int_a^b f \, d\alpha \right| \leq \int_a^b |f| \, d\alpha$$

2.49   Let $b_n = 1/(n+1)$ and let $\epsilon = 0.01$.

    a) Find an $N \in \mathbb{N}$ so that, for all $n > N$, then $|b_n - L| < \epsilon$ where $L = 0$.

    b) Find an $N \in \mathbb{N}$ so that, for all $n > N$, then $|b_n - L| < \epsilon$ where $L = 0.001$

    c) Explain why this does not contradict the uniqueness of limits.

2.50   Suppose that $a_n \to L > 0$. Show there is an $N \in \mathbb{N}$ so that $L/2 < a_n < 3L/2$ for all $n > N$.

2.51   Prove that a decreasing sequence that is bounded below converges.

2.52   Prove or disprove and salvage:

    a) $a_n \to 0$ iff $|a_n| \to 0$

    b) $a_n \to L$ iff $|a_n| \to |L|$

2.53   If there is a constant $K \in (0, 1)$ such that

$$|a_{n+2} - a_{n+1}| \leq K|a_{n+1} - a_n|$$

for all $n > N$, then $\{a_n\}$ is called *contractive*. Prove: every contractive sequence is Cauchy.

2.54   Write the repeating decimal $0.\bar{9}$ as a geometric series. Find the sum of the series.

2.55   Prove Theorem 2.58, the Cauchy criterion for series.

2.56   State and prove the analogue of Theorem 2.61, the limit comparison test, for the case

    a) $L = 0$

    b) $L = \infty$

2.57   Test for convergence or divergence.

    a) $\displaystyle\sum_{k=1}^\infty \frac{\ln(k)}{2k^3 - 1}$

    b) $\displaystyle\sum_{k=1}^\infty \frac{k}{3^{2k}}$

2.58   Show the series

$$\sum_{n=2}^\infty \left(1 - \frac{1}{n} \right)^n$$

is divergent.

2.59   Prove the *alternating series test*. If $a_n \geq 0$ for all n and $a_n \to 0$, then $\sum_{k=1}^\infty a_n$ converges.

2.60   If $\sum a_n$ converges but $\sum a_n^2$ diverges, then $\sum |a_n|$ diverges.

2.61   Describe *Raabe's test*.

2.62 Consider the series

$$\left[\frac{1}{2}\right] + \left[\frac{1 \cdot 3}{2 \cdot 4}\right] + \left[\frac{1 \cdot 3 \cdot 5}{2 \cdot 4 \cdot 6}\right] + \cdots$$

Show the series

    a) fails the ratio test

    b) fails the root test

    c) diverges by Raabe's test.

2.63 Does the following graph illustrate uniform convergence? Why or why not?

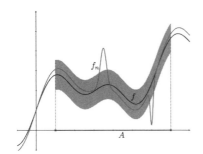

2.64 Determine whether $f_n(x) = x^n$ converges

    a) pointwise on $[0, 1]$.

    b) uniformly on $[0, 1]$.

    c) pointwise on $[1, 2]$.

    d) uniformly on $[0, 99/100]$.

2.65 Determine whether

$$g_n(x) = \frac{\sin(x)^n}{\ln(n + 1)}$$

converges

    a) pointwise on $[0, \pi]$.

    b) uniformly on $[0, \pi]$.

2.66 Let

$$f_n(x) = \frac{\sin(n^2 x)}{n}$$

on $\mathbb{R}$.

    a) Does f_n converge uniformly?

    b) Does f_n' converge uniformly?

2.67 Prove *Dini's theorem:* if f_n is a sequence of monotonic, continuous functions converging pointwise on a closed and bounded set F, then f_n converges uniformly on F.

2.68 Set

$$f_n(x) = nx(1 - x^2)^n$$

on $[0, 1]$. Does

$$\lim_n \int_0^1 f_n(x)\, dx = \int_0^1 \lim_n f_n(x)\, dx$$

2.69 Prove Theorem 2.69, the Weierstrass M-test.

2.70 Determine a power series representing $f(x) = \ln(1 + x)$.

2.71 Find the radius of convergence for the given power series.

    a) $\displaystyle\sum_{n=2}^{\infty} \left(\frac{x}{\ln(n)}\right)^n$

    b) $\displaystyle\sum_{n=0}^{\infty} \frac{x^n}{n! + n}$

2.72 Find the first 10 terms of the Taylor series for $\tan(x)$. Plot against $\tan(x)$. What do you observe?

2.73 Use Taylor series and integrability of power series to show

$$\int_0^x \ln(1+t)\, dt = \frac{x^2}{1 \cdot 2} - \frac{x^3}{2 \cdot 3} + \frac{x^4}{3 \cdot 4} - \cdots$$

for $|x| \leq 1$.

2.74 Who said,

If only I had the theorems! Then I should find the proofs easily enough.

INTERLUDE: EULER AND THE "BASEL PROBLEM"

In 1650, the Italian mathematician Pietro Mengoli published the text *Novæ quadraturæ arithmeticæ*. In this text, Mengoli analyzed geometric series and also showed that the harmonic series diverges (which was first proved by Oresme in 1350). Mengoli then went on to find exact values for series of the form $\sum 1/(n(n+r))$ for r up to 10. Quite a feat with only quill and parchment! Mengoli even demonstrated the sum $\sum 1/(n(n+1)(n+2)) = 1/4$. Nevertheless, he was stymied by the series $\sum 1/n^2$. [See Dunham (1999).]

The Swiss mathematician Jakob Bernoulli published *Tractatus siriebus infinitis* (1689) on infinite series. The title page appears in Figure A.1. In *Tractatus*, Bernoulli showed that the series $\sum 1/n^2$ converged by comparing the terms

$$\frac{1}{n} - \frac{1}{n+1} = \frac{1}{n(n+1)} \leq \frac{1}{n^2} \leq \frac{2}{n(n+1)} = \frac{2}{n} - \frac{2}{n+1}$$

Bernoulli then knew that

$$1 = \sum_{n=1}^{\infty} \frac{1}{n(n+1)} \leq \sum_{n=1}^{\infty} \frac{1}{n^2} \leq \sum_{n=1}^{\infty} \frac{2}{n(n+1)} = 2$$

However, Bernoulli couldn't find the exact sum writing that it "is more difficult than one would expect." Many other top mathematicians also failed; for instance, Leibniz wasn't able to find the exact value, either. Bernoulli wrote in the *Tractatus*

> If anyone finds and communicates to us that which thus far has eluded our efforts, great will be our gratitude.

As the *Tractatus* was published in Basel where Bernoulli lived and was a professor of mathematics at the university, finding the exact value of the series became known as the *Basel problem*. The *Tractatus* was republished posthumously as an Appendix in *Ars Conjectandi*.

Before we look at Euler's answer to the Basel problem, let's consider a tantalizing experiment that Bernoulli likely had done.

An Experiment

Consider the two series

$$B_N = \sum_{n=2}^{N} \frac{1}{n^2} \quad \text{and} \quad C_N = \sum_{n=2}^{N} \frac{1}{n^2 - 1}$$

Note that we start both summations at $n = 2$.

1. Determine which is larger: $1/n^2$ or $1/(n^2 - 1)$.

2. Examine the errors $e_n = 1/(n^2 - 1) - 1/n^2$ and $E_N = C_N - B_N$.

JACOBI BERNOULLI,
Profeſſ. Baſil. & utriuſque Societ. Reg. Scientiar.
Gall. & Pruſſ. Sodal.
MATHEMATICI CELEBERRIMI,

ARS CONJECTANDI,

OPUS POSTHUMUM.

Accedit

TRACTATUS
DE SERIEBUS INFINITIS,

Et EPISTOLA Gallicè ſcripta

DE LUDO PILÆ
RETICULARIS.

BASILEÆ,
Impenſis THURNISIORUM, Fratrum.
cIↄ Iↄcc xiii.

Figure A.1   Title Page of Jakob Bernoulli's *Ars Conjectandi*

3. Write the partial fraction expansion of $1/(n^2 - 1)$.

4. Find the exact sum of the series $\sum_{n=2}^{\infty} 1/(n^2 - 1)$.

5. Estimate the answer to the Basel problem.

We are so close, but shifted by 1. The answer still eludes us.

Euler's Ingenious Solution

In 1731, Euler found a very clever approximating technique based on two different series expansions of the integral $\int_0^{1/2} -\ln(1-t)/t \, dt$ [Dunham (1999), p. 43]. Then, in his 1735 paper "On the sums of series of reciprocals,"[6] Euler, not yet 30 years old, solved the Basel problem, showing the answer to be surprisingly $\pi^2/6$. His method refined the approach that led to the 1731 approximation. Let's look at Euler's argument.

Theorem I.1 (The Basel Problem)

$$\sum_{n=1}^{\infty} \frac{1}{n^2} = \frac{\pi^2}{6}$$

[6]Originally published as "De summis serierum reciprocarum" in *Commentarii academiae scientiarum Petropolitanae* 7, 1740, pp. 123–134; reprinted in *Opera Omnia*: Series 1, Volume 14, pp. 73–86.

Proof: Define the series $S(x) = (x - x^3/3! + x^5/5! - x^7/7! \pm \cdots)/x$ which is the series expansion of $\sin(x)/x$. Assume $x \neq 0$. Then a root $\hat{x}$ of $S(x)$ is a nonzero root of $\sin(x)$. Therefore $\hat{x} = \pm k\pi$ for $k \in \mathbb{N}$. When $x = 0$, we have that $\lim_{x \to 0} S(x) = 1$. By the analogue of the factor theorem for polynomials, we can write S in the form

$$S(x) = \left(1 - \frac{x}{\pi}\right)\left(1 + \frac{x}{\pi}\right)\left(1 - \frac{x}{2\pi}\right)\left(1 + \frac{x}{2\pi}\right)\left(1 - \frac{x}{3\pi}\right)\left(1 + \frac{x}{3\pi}\right)\cdots$$

Combine the factors in pairs.

$$S(x) = \left(1 - \frac{x^2}{\pi^2}\right)\left(1 - \frac{x^2}{2^2\pi^2}\right)\left(1 - \frac{x^2}{3^2\pi^2}\right)\cdots$$

Calculate the coefficient of x^2 in both expansions—the coefficients must be equal!

$$-\frac{1}{3!} = -\left(\frac{1}{\pi^2} + \frac{1}{4\pi^2} + \frac{1}{9\pi^2} + \frac{1}{16\pi^2} + \cdots\right)$$

Multiplying both sides by $-\pi^2$ solves the Basel problem.   ■

Euler recognized that there were logical gaps in his argument (*Find them!*), subsequently publishing other derivations. [See Dunham (1999, p. 55–60); also see Aigner & Ziegler (2001, p. 29–33) for other elegant proofs.] Euler also recognized his technique gave further sums; he calculated exact values for $\sum 1/k^p$ for $p = 4, 6, 8$, all the way to $p = 26$, showing each to be a rational multiple of π^p. It is still unknown whether $\sum 1/k^3$ is a rational multiple of π^3.

The solution to the Basel problem marked the beginning of Euler's illustrious and prodigious career. Euler produced a huge number of papers, well over 800, and even created new fields of mathematics such as graph theory. The best advice comes from Pierre-Simon Laplace[7]:

> "Read Euler, read Euler, he is the master of us all."

Perhaps the best way for us to start today is to carefully read William Dunham's text *Euler, the Master of Us All.*

---

[7]Count G. Libri wrote in the *Journal des Savants*, January 1846, p. 51, of Laplace: "ces paroles mémorables que nous avons entendues de sa propre bouche: 'Lisez Euler, lisez Euler, c'est notre maître à tous'."

CHAPTER 3

A BRIEF INTRODUCTION TO LEBESGUE THEORY

Introduction

The span from Newton and Leibniz to Lebesgue covers only 250 years (Figure 3.1). Lebesgue published his dissertation "Intégrale, longueur, aire" ("Integral, length, area") in the *Annali di Matematica* in 1902. Lebesgue developed "measure of a set" in the first chapter and an integral based on his measure in the second.

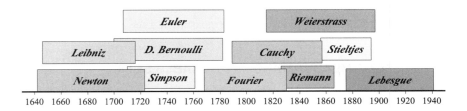

Figure 3.1   From Newton and Leibniz to Lebesgue.

Introduction to Real Analysis. By William C. Bauldry
Copyright © 2009 John Wiley & Sons, Inc.

Part of Lebesgue's motivation were two problems that had arisen with Riemann's integral. First, there were functions for which the integral of the derivative does not recover the original function and others for which the derivative of the integral is not the original. Second, the integral of the limit of a sequence of functions was not necessarily the limit of the integrals. We've seen that uniform convergence allows the interchange of limit and integral, but there are sequences that do not converge uniformly yet the limit of the integrals is equal to the integral of the limit. In Lebesgue's own words from "Integral, length, area" (as quoted by Hochkirchen (2004, p. 272)),

> It thus seems to be natural to search for a definition of the integral which makes integration the inverse operation of differentiation in as large a range as possible.

Lebesgue was able to combine Darboux's work on defining the Riemann integral with Borel's research on the "content" of a set. Darboux was interested in the interplay of the definition of integral with discontinuous functions and in the convergence problems. It was Darboux's development of the Riemann integral that we followed in Chapter 2. Borel (who was Lebesgue's thesis advisor) needed to describe the size of sets of points on which a series converged; he expanded on Jordan's definition of the content of a set which itself was an expansion of Peano's definition of content measuring the size of a set. Peano's work was motivated by Hankel's attempts to describe the size of the set of discontinuities of a Riemann integrable function and by an attempt to define integration analytically, as opposed to geometrically (Hawkins, 2002, chapter 4). Rarely, if ever, is revolutionary mathematics done in isolation. Hochkirchen's (2004) essay "The Theory of Measure and Integration from Riemann to Lebesgue" gives a detailed historical perspective.

Another problem also provided primary motivation for Lebesgue: the question of convergence and integrating series term by term. Newton had used series expansions cleverly to integrate functions when developing calculus. Fourier thought that it was always valid to integrate a trigonometric series representation of a function term by term. Cauchy believed continuity of the terms sufficed; Cauchy's integral required continuity to exist. Then Abel gave an example that didn't work. Weierstrass recognized that uniform convergence was the key to term-by-term integration. Dirichlet developed wildly discontinuous counterexamples such as his "monster." Riemann defined his integral so as to not require continuity, but uniform convergence of the series was still necessary for term-by-term integration. However, some nonuniformly convergent series could still be integrated term by term. What is the right condition? Lebesgue's theory can answer these questions.

We now turn to Lebesgue's concept of the measure of a set.

3.1 LEBESGUE MEASURE AND MEASURABLE SETS

Lebesgue's measure is an extended real-valued set function, a function from a collection of sets into $[0, \infty]$. Measure is based on the lengths of open intervals as these intervals are the basic building blocks of open sets in the reals. The best measure μ would satisfy four properties:

1. For each interval I, $\mu(I) = \text{length}(I)$.

2. For $x \in \mathbb{R}$ and $E \subset \mathbb{R}$, $\mu(x + E) = \mu(E)$.

3. For a sequence of disjoint sets $\{E_n\}$, $\mu\left(\bigcup_n E_n\right) = \sum_n \mu(E_n)$.

4. Every subset of $\mathbb{R}$ can be measured by μ.

Unfortunately, we're asking too much; it's not possible to satisfy all four properties. Property 4 is the first to fall. There will be sets that cannot be measured. (Look ahead to "Vitali's Nonmeasurable Set," p. 249.) If we assume the *continuum hypothesis* (see p. 172) is true, then we cannot have a measure satisfying properties 1, 3, and 4. We are unwilling to give up the first two; we'll weaken the third initially, hoping to recapture it later.

Algebras and σ-Algebras of Sets

An *algebra of sets* is a nonempty collection that behaves well with respect to unions and complements. A collection $\mathcal{A}$ of sets is an *algebra* if and only if the union of any two sets in $\mathcal{A}$ is also in $\mathcal{A}$, and the complement of every set in $\mathcal{A}$ is also in $\mathcal{A}$. De Morgan's law then implies the intersection of any two sets in $\mathcal{A}$ is in $\mathcal{A}$. By extension, any union or intersection of a finite number of sets in $\mathcal{A}$ is also in $\mathcal{A}$. De Morgan's laws give us a principle of duality for statements about sets in an algebra: replace unions with intersections and vice versa to obtain a new true statement.

Suppose that $A \neq \emptyset$ is a member of an algebra $\mathcal{A}$ of subsets of X. Then $A^c \in \mathcal{A}$. Since A and A^c are in $\mathcal{A}$, we have that $A \cup A^c = X$ and $A \cap A^c = \emptyset$ are both members of $\mathcal{A}$. In particular, every algebra of sets of reals contains both $\emptyset$ and $\mathbb{R}$.

An algebra of sets $\mathcal{A}$ is a σ-*algebra* if every union of a countable collection of sets from $\mathcal{A}$ is also in $\mathcal{A}$. Once again, De Morgan's laws tell us that σ-algebras are closed with respect to countable intersections.

■ EXAMPLE 3.1

1. For any $X \neq \emptyset$, the collection $\{\emptyset, X\}$ forms a σ-algebra, the *trivial σ-algebra*. *Prove this!*

2. Let $\mathcal{P}(\mathbb{N})$ be the *power set* of $\mathbb{N}$, the set of all subsets of $\mathbb{N}$. Then $\mathcal{P}(\mathbb{N})$ is a σ-algebra. *Show this!*

3. Let $\mathcal{F}$ be the collection of subsets of an infinite set X that are finite or have finite complement, the *co-finite algebra*.

   (a) $\mathcal{F}$ is an algebra.

   The complement of any set in $\mathcal{F}$ is clearly in $\mathcal{F}$. *Why?*

   Since the union of two finite sets is finite, the union of two sets with finite complement has finite complement, and the union of a finite set with a set having finite complement has finite complement, $\mathcal{F}$ is closed under unions.

Hence $\mathcal{F}$ is an algebra.

(b) $\mathcal{F}$ is *not* a σ-algebra.

Since X is infinite, we can choose a subset $Y \subset X$ such that Y is countable and Y^c is infinite. For each $y \in Y$, the set $\{y\}$ is finite, and hence in $\mathcal{F}$. However, the union $\bigcup_{y \in Y}\{y\} = Y$, which is not a member of $\mathcal{F}$. *Give an example of such a Y when $X = \mathbb{N}$.*

4. The *Borel σ-algebra on* $\mathbb{R}$ is the smallest σ-algebra containing all the open sets and is denoted by $\mathcal{B}(\mathbb{R})$.

∎

Lebesgue Outer Measure

A collection of open intervals $\{I_n \mid n = 1, 2, \dots\}$ *covers* a set E if $E \subseteq \bigcup_n I_n$. Since the intervals are open, we call $\{I_n\}$ an *open cover* of E. Define the *length* ℓ of the open interval $I = (a, b)$ to be $\ell(I) = b - a$. We combine open covers and length to measure the size of a set. Since the cover contains the set, we'll call it the *outer measure*. The outer measure is extremely close to the measure Jordan defined in 1892.

Definition 3.1 (Lebesgue Outer Measure) *For any set $E \subseteq \mathbb{R}$, define the* Lebesgue outer measure μ^* *of E to be*

$$\mu^*(E) = \inf_{E \subset \bigcup I_n} \sum_n \ell(I_n)$$

the infimum of the sums of the lengths of open covers of E.

If $A \subseteq B$, then any open cover of B also covers A. Therefore $\mu^*(A) \leq \mu^*(B)$. This property is called *monotonicity*.

Theorem 3.1 *The outer measure of an interval is its length.*

Proof: First, consider a bounded, closed interval $I = [a, b]$. For any $\epsilon > 0$,

$$[a, b] \subset (a - \epsilon/2, b + \epsilon/2)$$

Hence, $\mu^*(I) \leq b - a + \epsilon$. Since $\epsilon > 0$ is arbitrary, $\mu^*([a, b]) \leq b - a$.

Now let $\{I_n\}$ be an open cover of $[a, b]$. The Heine-Borel theorem states that since $[a, b]$ is closed and bounded there is a finite subcover $\{I_k \mid k = 1..N\}$ for I. Order the intervals so they overlap, starting with the first containing a and ending with the last containing b. *How do we know we can do this?* Thus

$$\sum_{k=1}^{N} \ell(I_k) = (b_1 - a_1) + (b_2 - a_2) + \cdots + (b_N - a_N)$$

$$= b_N - (a_N - b_{N-1}) - (a_{N-1} - b_{N-2}) - \cdots - (a_2 - b_1) - a_1$$

$$\geq b_N - a_1$$

$$> b - a$$

Thus $\mu^*(I) \geq b - a$, which combines with the first inequality to yield $\mu^*(I) = b - a$.

Second, let I be any bounded interval and let $\epsilon > 0$. There is a closed interval $J \subset I$ such that $\ell(I) - \epsilon < \ell(J)$. Then

$$\ell(I) - \epsilon < \ell(J) = \mu^*(J) \leq \mu^*(I) \leq \mu^*(\bar{I}) = \ell(\bar{I}) = \ell(I)$$

or

$$\ell(I) - \epsilon < \mu^*(I) \leq \ell(I).$$

Since $\epsilon > 0$ is arbitrary, we have $\mu^*(I) = \ell(I)$.

Last, if I is an infinite interval, for each $n \in \mathbb{N}$, there is a closed interval $J \subset I$ with $\mu^*(J) = n$. Then $n = \mu^*(J) \leq \mu^*(I)$ implies that $\mu^*(I) = \infty$. ∎

While we can't satisfy additivity (property 3), we can weaken the property by changing the equality to less than or equal. This relation is called *subadditivity*.

Theorem 3.2 *Lebesgue outer measure is countably subadditive. Let $\{E_n\}$ be a countable sequence of subsets of $\mathbb{R}$. Then*

$$\mu^*\left(\bigcup_n E_n\right) \leq \sum_n \mu^*(E_n)$$

Proof: If any set E_n has infinite outer measure, the inequality is trivially true. Suppose that $\mu^*(E_n) < \infty$ for all n and let $\epsilon > 0$. For each n, there is a countable collection of open intervals $\{I_{n,j} \mid j \in \mathbb{N}\}$ that covers E_n and such that

$$\sum_{j \in \mathbb{N}} \ell(I_{n,j}) < \mu^*(E_n) + \frac{\epsilon}{2^n}$$

A countable collection of countable sets is countable, so $\{I_{n,j} \mid n, j \in \mathbb{N}\}$ is a countable, open cover of $E = \bigcup E_n$. Therefore

$$\mu^*\left(\bigcup_n E_n\right) \leq \sum_{n,j} \ell(I_{n,j}) = \sum_{n \in \mathbb{N}} \left(\sum_{j \in \mathbb{N}} \ell(I_{n,j})\right)$$

$$< \sum_{n \in \mathbb{N}} \left(\mu^*(E_n) + \frac{\epsilon}{2^n}\right)$$

$$= \sum_{n \in \mathbb{N}} \mu^*(E_n) + \epsilon$$

Thus $\mu^*\left(\bigcup_n E_n\right) \leq \sum_n \mu^*(E_n)$. ∎

Several results are immediate corollaries.

Corollary 3.3 *The outer measure of a countable set is zero.*

We know that $\mu^*([0, 1]) = 1$.

Corollary 3.4 *The interval $[0, 1]$ is not countable.*

The union of a set of open intervals is an open set. The definition of infimum tells us that there is an open cover with outer measure within ϵ of the set it covers. Combine these observations to find an open set with outer measure arbitrarily close to the outer measure of a given set.

Corollary 3.5 *Let $E \subseteq \mathbb{R}$ and $\epsilon > 0$. There is an open set O such that $E \subseteq O$ and $\mu^*(E) \leq \mu^*(O) \leq \mu^*(E) + \epsilon$.*

With the outer measure in hand, we ask, "What is a measurable set?"

Lebesgue Measure

In 1902, Lebesgue defined the *inner measure* of a set in terms of the outer measure of the complement of the set. If the outer and inner measures were equal, Lebesgue defined the measure of the set to be that common value. One drawback to this approach is that it can be difficult unless the measure of the whole space is finite. [See, e.g., Hochkirchen (2004).] As a way to generalize the concept, Carathéodory developed a method of constructing a measure using only the outer measure. We'll follow Carathéodory's lead. First, we determine which sets are measurable, then we define the Lebesgue measure for these sets.

Definition 3.2 (Carathéodory's Condition) *A set E is* Lebesgue measurable *if and only if for every set $A \subseteq \mathbb{R}$ we have*

$$\mu^*(A) = \mu^*(A \cap E) + \mu^*(A \cap E^c)$$

Let $\mathfrak{M}$ be the family of all Lebesgue measurable sets.

Informally, a set is measurable if it splits every other set into two pieces with measures that add correctly. The definition of measurable is symmetric: if E is measurable, so is E^c; i.e., if $E \in \mathfrak{M}$, then $E^c \in \mathfrak{M}$. Also, it is easily seen that $\emptyset$ and $\mathbb{R} \in \mathfrak{M}$.

Theorem 3.6 *If $\mu^*(E) = 0$, then E is measurable.*

Proof: For any set A, it's true that

$$\mu^*(A) = \mu^*((A \cap E) \cup (A \cap E^c)) \leq \mu^*(A \cap E) + \mu^*(A \cap E^c)$$

Since $A \cap E \subseteq E$, we see that $\mu^*(A \cap E) \leq \mu^*(E) = 0$. Thus $\mu^*(A \cap E) = 0$. Now note that $A \cap E^c \subseteq A$, so $\mu^*(A \cap E^c) \leq \mu^*(A)$. Hence

$$\mu^*(A) \leq \mu^*(A \cap E) + \mu^*(A \cap E^c) \leq \mu^*(A)$$

Thus $E \in \mathfrak{M}$. ∎

If two sets are measurable, is their union also measurable?

Theorem 3.7 *The union of two measurable sets is measurable.*

Proof: Let E_1 and E_2 be two measurable sets. Let A be a set to use in Carathédory's condition. Apply Carathédory's condition to the set $A \cap E_1^c$ to have

$$\mu^*(A \cap E_1^c) = \mu^*((A \cap E_1^c) \cap E_2) + \mu^*((A \cap E_1^c) \cap E_2^c)$$

Now

$$A \cap (E_1 \cup E_2) = (A \cap E_1) \cup (A \cap E_2)$$
$$= (A \cap E_1) \cup (A \cap E_2 \cap E_1^c)$$

so

$$\mu^*(A \cap (E_1 \cup E_2)) + \mu^*(A \cap (E_1 \cup E_2)^c)$$
$$= \mu^*(A \cap (E_1 \cup E_2)) + \mu^*(A \cap (E_1^c \cap E_2^c))$$
$$\leq [\mu^*(A \cap E_1) + \mu^*(A \cap E_2 \cap E_1^c)] + \mu^*(A \cap E_1^c \cap E_2^c)$$
$$= \mu^*(A \cap E_1) + [\mu^*(A \cap E_2 \cap E_1^c) + \mu^*(A \cap E_1^c \cap E_2^c)]$$
$$= \mu^*(A \cap E_1) + [\mu^*((A \cap E_1^c) \cap E_2) + \mu^*((A \cap E_1^c) \cap E_2^c)]$$

Since E_2 is measurable, then

$$\mu^*(A \cap (E_1 \cup E_2)) + \mu^*(A \cap (E_1 \cup E_2)^c) = \mu^*(A \cap E_1) + \mu^*(A \cap E_1^c)$$

Now, since E_1 is measurable, we have

$$\mu^*(A \cap (E_1 \cup E_2)) + \mu^*(A \cap (E_1 \cup E_2)^c) = \mu^*(A)$$

Therefore
$$\mu^*(A) = \mu^*(A \cap (E_1 \cup E_2)) + \mu^*(A \cap (E_1 \cup E_2)^c)$$

which yields $E_1 \cup E_2 \in \mathfrak{M}$. ∎

The results above indicate that the collection of measurable sets $\mathfrak{M}$ forms an algebra.

Theorem 3.8 *A countable union of measurable sets is measurable.*

Proof: Let $\{E_k\}$ be a countable sequence of measurable sets and put $E = \bigcup E_k$. Let A be an abitrary test set to use in Carathédory's condition. By subadditivity,

$$\mu^*(A) \leq \mu^*((A \cap E) + \mu^*(A \cap E^c)$$

Set $F_n = \bigcup_{k=1}^n E_k$ and $F = \bigcup_{k=1}^\infty E_k = E$. Also, set $G_1 = E_1, G_2 = E_2 - E_1$, $G_3 = E_3 - E_1 - E_2$, etc., and $G = \bigcup_{k=1}^\infty G_k$. Then

1. $G_i \cap G_j = \emptyset$ for $i \neq j$ (G_k *are pairwise disjoint*)

2. $F_n = \bigcup\limits_{k=1}^{n} G_k$

3. $G = F = E$

Each F_n and G_n are measurable. *Verify this!* Therefore,

$$\mu^*(A) = \mu^*(A \cap F_n) + \mu^*(A \cap F_n^c)$$

Apply Carathédory's condition to $A \cap F_n$ to see

$$\mu^*(A \cap F_n) = \mu^*((A \cap F_n) \cap G_n) + \mu^*((A \cap F_n) \cap G_n^c)$$
$$= \mu^*(A \cap G_n) + \mu^*(A \cap F_{n-1})$$

Why? Iterate this relation to obtain

$$\mu^*(A \cap F_n) = \sum_{k=1}^{n} \mu^*(A \cap G_k)$$

Since $F_n \subseteq F$, then $F^c \subseteq F_n^c$ for each n, so $\mu^*(A \cap F_n^c) \geq \mu^*(A \cap F^c)$. Hence

$$\mu^*(A) \geq \sum_{k=1}^{n} \mu^*(A \cap G_k) + \mu^*(A \cap F^c)$$

The summation above is increasing and bounded (*Why?*), and therefore convergent. So

$$\mu^*(A) \geq \sum_{k=1}^{\infty} \mu^*(A \cap G_k) + \mu^*(A \cap F^c)$$

But

$$\sum_{k=1}^{\infty} \mu^*(A \cap G_k) \geq \mu^* \left(\bigcup_{k=1}^{\infty} (A \cap G_k) \right) = \mu^* \left(A \cap \bigcup_{k=1}^{\infty} G_k \right) = \mu^*(A \cap F)$$

So we have

$$\mu^*(A) \geq \mu^*(A \cap F) + \mu^*(A \cap F^c)$$

Since we have already shown the reverse inequality, then

$$\mu^*(A) = \mu^*(A \cap F) + \mu^*(A \cap F^c)$$

Therefore, by Carathédory's condition, $F = \bigcup E_k \in \mathfrak{M}$. ∎

We have shown:

Corollary 3.9 *The collection of measurable sets $\mathfrak{M}$ is a σ-algebra.*

Corollary 3.10 *The Borel sets are measurable.*

Proof: Since $\mathcal{B}(\mathbb{R})$ is the smallest σ-algebra containing all the open sets and $\mathfrak{M}$ is a σ-algebra containing all open intervals, and hence all open sets, then $\mathcal{B}(\mathbb{R}) \subseteq \mathfrak{M}$. ∎

There are measurable sets that are not Borel sets. The construction of a measurable, non-Borel set is beyond our scope; see Cohn (1980, p. 56) for details.

Even though $\mathfrak{M}$ is huge, there must be sets that are not measurable. So $\mathfrak{M} \subsetneq \mathcal{P}(\mathbb{R})$. Vitali (1905) constructed the first example of a nonmeasurable set. (Look ahead to "Vitali's Nonmeasurable Set," p. 249.) Vitali's construction leads to the result that any translation-invariant measure (Exercise 3.8) has nonmeasurable sets.

Now that we've studied the class of measurable sets $\mathfrak{M}$, we are ready to define Lebesgue measure.

Definition 3.3 (Lebesgue Measure) *The* Lebesgue measure μ *is the restriction of the outer measure* μ^* *to the measurable sets* $\mathfrak{M}$. *That is, for* $E \in \mathfrak{M}$, *set* $\mu(E) = \mu^*(E)$.

■ **EXAMPLE 3.2**

Any set E with outer measure zero is measurable and so is Lebesgue measurable with $\mu(E) = 0$.

Definition 3.4 (Almost Everywhere) *A property is said to hold* almost everywhere *if and only if the measure of the set on which the property does not hold is zero.*

■ **EXAMPLE 3.3**

Let $\chi_{\mathbb{N}}(x)$ be the characteristic function of the integers, that is,

$$\chi_{\mathbb{N}}(x) = \begin{cases} 1 & x \in \mathbb{N} \\ 0 & \text{otherwise} \end{cases}$$

Since $\mu(\mathbb{N}) = 0$ (*Show this!*), then $\chi_{\mathbb{N}} = 0$ almost everywhere. ■

The term almost everywhere is often abbreviated "a.e." Is $\chi_{\mathbb{N}}$ continuous a.e.?

Outer measure is countably subadditive; therefore Lebesgue measure is countably subadditive, too. However, if we have a sequence of pairwise disjoint sets, we can say more. We can recapture property 3, additivity!

Theorem 3.11 (Additivity of Lebesgue Measure) *Let* $\{E_n\}$ *be a countable sequence of pairwise disjoint, measurable sets. Then*

$$\mu\left(\bigcup_{n=1}^{\infty} E_n\right) = \sum_{n=1}^{\infty} \mu(E_n)$$

Proof: If $\{E_n\}$ is a finite sequence, we'll use induction on n, the number of sets in the sequence. For $n = 1$, the result trivially holds. Assume the equation holds for $n - 1$ sets. Since the sets are disjoint,

$$\left(\bigcup_{k=1}^{n} E_k \right) \cap E_n = E_n \quad \text{and} \quad \left(\bigcup_{k=1}^{n} E_k \right) \cap E_n^c = \bigcup_{k=1}^{n-1} E_k$$

A finite union of measurable sets is measurable and E_n is measurable; thus

$$\mu \left(\bigcup_{k=1}^{n} E_k \right) = \mu \left(\left[\bigcup_{k=1}^{n} E_k \right] \cap E_n \right) + \mu \left(\left[\bigcup_{k=1}^{n} E_k \right] \cap E_n^c \right)$$

$$= \mu \left(E_n \right) + \mu \left(\bigcup_{k=1}^{n-1} E_k \right)$$

$$= \mu \left(E_n \right) + \sum_{k=1}^{n-1} \mu(E_k)$$

The result follows by induction.

If $\{E_n\}$ is an infinite sequence, then $\bigcup_{k=1}^{n} E_k \subset \bigcup_{k=1}^{\infty} E_k$. Hence

$$\mu \left(\bigcup_{k=1}^{\infty} E_k \right) \geq \mu \left(\bigcup_{k=1}^{n} E_k \right) = \sum_{k=1}^{n} \mu(E_k)$$

We have a bounded, increasing series on the right. Thus

$$\mu \left(\bigcup_{k=1}^{\infty} E_k \right) \geq \sum_{k=1}^{\infty} \mu(E_k)$$

Countable subadditivity supplies the reverse inequality, so

$$\mu \left(\bigcup_{k=1}^{\infty} E_k \right) = \sum_{k=1}^{\infty} \mu(E_k)$$

Hence the result holds for finite or countably infinite sequences of disjoint sets. ∎

■ **EXAMPLE 3.4**

Let $a \in \mathbb{R}$ and set $I = (a, \infty)$. Show $\mu(I) = \infty$.

We already know that $\mu((a, \infty)) = \infty$ is true, but we'll use additivity for a different verification. Define $E_k = (a + k - 1, a + k]$ for $k = 1, 2, \ldots$. Then $\mu(E_k) = 1$ for all k. Further, $E_i \cap E_j = \emptyset$ for $i \neq j$; that is, the sequence is pairwise disjoint. So

$$\mu((a, \infty)) = \mu \left(\bigcup_{k=1}^{\infty} E_k \right) = \sum_{k=1}^{\infty} \mu(E_k) = \sum_{k=1}^{\infty} 1 = \infty$$

■

If we have a decreasing sequence of nested sets, then we can calculate the measure of the intersection as the limit of the measures. In a certain sense, this statement says "the measure of the limit is the limit of measures" for nested sets. This result will be quite useful to us in integration.

Theorem 3.12 *Let* $\{E_n\}$ *be an infinite sequence of nested, measurable sets, that is, for each* n, $E_{n+1} \subset E_n$. *If* $\mu(E_1)$ *is finite, then*

$$\mu \left(\bigcap_{k=1}^{\infty} E_k \right) = \lim_{n \to \infty} \mu(E_n)$$

Proof: Let $E = \bigcap_k E_k$ and define $F_k = E_k - E_{k+1}$. The F_k are pairwise disjoint because the E_k are nested, and

$$\bigcup_{k=1}^{\infty} F_k = E_1 - E$$

Since E and E_1 are measurable, so is $E_1 - E$. Then

$$\mu(E_1 - E) = \mu \left(\bigcup_{k=1}^{\infty} F_k \right) = \sum_{k=1}^{\infty} \mu(F_k)$$

By the definition of F_k, we have

$$\mu(E_1 - E) = \sum_{k=1}^{\infty} \mu(E_k - E_{k+1})$$

Since $E_1 \supset E_2 \supset E_3 \supset \cdots \supset E$ and $\mu(E_1) < \infty$, the measure of all E_k and of E is finite. Thus, $\mu(E_1) = \mu(E) + \mu(E_1 - E)$ (*Why?*) implies that $\mu(E_1 - E) = \mu(E_1) - \mu(E)$. Continuing, $\mu(E_k) = \mu(E_{k+1}) + \mu(E_k - E_{k+1})$ implies that $\mu(E_k - E_{k+1}) = \mu(E_k) - \mu(E_{k+1})$ for each k. Substituting, we see

$$\mu(E_1) - \mu(E) = \sum_{k=1}^{\infty} \mu(E_k) - \mu(E_{k+1})$$

$$= \lim_{n \to \infty} \sum_{k=1}^{n} \mu(E_k) - \mu(E_{k+1})$$

$$= \lim_{n \to \infty} (\mu(E_1) - \mu(E_{k+1}))$$

$$= \mu(E_1) - \lim_{n \to \infty} \mu(E_{k+1})$$

Since $\mu(E_1) < \infty$, the result holds. ∎

■ **EXAMPLE 3.5   The Cantor Set**

We will define the *Cantor set* C as the infinite intersection of a nested sequence
of sets. Set $C_0 = [0, 1]$ and note that $\mu(C_0) = 1$. Now define C_1 to be C_0 minus
the open middle third interval; $C_1 = [0, 1/3] \cup [2/3, 1]$. Note $\mu(C_1) = 2/3$.
Continue the process, removing the middle third of each interval, to have

$$C_2 = \left[0, \frac{1}{9}\right] \cup \left[\frac{2}{9}, \frac{1}{3}\right] \cup \left[\frac{2}{3}, \frac{7}{9}\right] \cup \left[\frac{7}{9}, 1\right]$$

$$C_3 = \left[0, \frac{1}{27}\right] \cup \left[\frac{2}{27}, \frac{3}{27}\right] \cup \left[\frac{6}{27}, \frac{7}{27}\right] \cup \left[\frac{8}{27}, \frac{9}{27}\right]$$

$$\cup \left[\frac{18}{27}, \frac{19}{27}\right] \cup \left[\frac{20}{27}, \frac{21}{27}\right] \cup \left[\frac{24}{27}, \frac{25}{27}\right] \cup \left[\frac{26}{27}, 1\right]$$

and so forth. Figure 3.2 shows several iterations of the procedure.

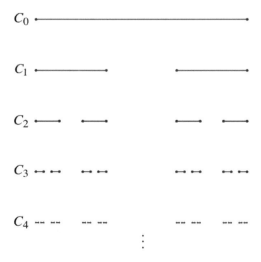

Figure 3.2   Constructing the Cantor Set

Each C_n is a finite union of closed intervals, and thus closed. For every n, $C_{n+1} \subset C_n$, and the measure of C_{n+1} is $2/3$ the measure of C_n. Hence

$$\mu(C_0) = 1$$

$$\mu(C_1) = \frac{2}{3} C_0 = \frac{2}{3}$$

$$\mu(C_2) = \frac{2}{3} C_1 = \left(\frac{2}{3}\right)^2$$

$$\vdots$$

$$\mu(C_n) = \frac{2}{3} C_{n-1} = \left(\frac{2}{3}\right)^n$$

$$\vdots$$

The Cantor set C is the intersection of the C_n's,

$$C = \bigcap_{n=0}^{\infty} C_n$$

Since we are removing the middles of the intervals, the endpoints remain in the sets, doubling the number of endpoints in each iteration. This doubling can be used to create a bijective map from C onto $\mathbb{R}$ showing that C is uncountable.

What is the measure of the Cantor set? By Theorem 3.12, we see

$$\mu(C) = \lim_{n \to \infty} \mu(C_n) = \lim_{n \to \infty} \left(\frac{2}{3}\right)^n = 0$$

The Cantor set is very interesting. The set C is a *perfect set* where every point is an accumulation point but the Cantor set contains no interval. Most fascinating for us, though, is the fact that the Cantor set is an uncountable set with measure zero! ∎

The property of being measurable is more general than being open or closed. Consider the interval $[0, 1)$. This interval is neither open nor closed but is easily seen to be measurable. How strange can a measurable set be? In one sense, not very.

Theorem 3.13 *Let $E \subseteq \mathbb{R}$. The following are equivalent.*

1. *$E \in \mathfrak{M}$; i.e., E is measurable,*

2. *To any $\epsilon > 0$, there is an open set $O \supset E$ such that $\mu^*(O - E) < \epsilon$,*

3. *To any $\epsilon > 0$, there is a close set $F \subset E$ such that $\mu^*(E - F) < \epsilon$.*

The proof is left to the exercises.

Define the *symmetric difference* $\triangle$ of two sets A and B by

$$A \triangle B = (A - B) \cup (B - A)$$

Theorem 3.14 *If $E \subset [a, b]$ is measurable, then, for any $\epsilon > 0$, there is a finite collection of open intervals $I_1, I_2, \ldots, I_n$ such that $\mu(E \triangle \bigcup I_n) < \epsilon$.*

Proof: Since E is measurable and any open set in $\mathbb{R}$ is the countable union of open intervals, there is a countable union of open intervals $U = \bigcup I_j$ such that

$$\mu(E) \leq \mu(U) \leq \mu(E) + \frac{\epsilon}{2}$$

Thus $\mu(U - E) < \epsilon/2$. *Why?*

Put $F = [a, b] - E$; then F is measurable. Similarly, there is a countable collection of open intervals V containing F with

$$\mu(V - F) < \frac{\epsilon}{2}$$

Since V is open, V^c is closed. Also,

$$V - F = V - ([a, b] - E) \supset V \cap E = E - V^c$$

Therefore,

$$\mu(E - V^c) \leq \mu(V - F) < \frac{\epsilon}{2}$$

Now, U is a countable collection of open intervals that covers the closed and bounded set $V^c \cap [a, b] \subseteq E$. By the Heine-Borel theorem, U has a finite subcover $W = \bigcup_{j=1}^{n} I_j$ that covers $V^c \cap [a, b]$. Then

$$\mu^*(E \triangle W) \leq \mu^*(E \cap W^c) + \mu^*(E^c \cap W)$$
$$\leq \mu^*(E - V^c) + \mu^*(U - E) < \epsilon$$

(*Why?*), and the result holds.   ∎

So a measurable set is "almost" an open set or a closed set. On the other hand, we know there are measurable sets that are not in $\mathcal{B}(\mathbb{R})$, that is, that are not Borel sets. Since $\mathcal{B}(\mathbb{R})$ is a σ-algebra that contains all the open sets, it also contains all countable unions of open sets (or closed sets), and so all countable intersections of countable unions of open sets, and so all countable unions of countable intersections of countable unions of open sets, and so forth. A measurable set that is not a Borel must have a very complicated structure. Nevertheless, a measurable set must be sandwiched between a closed set and an open set that are arbitrarily close to each other.

3.2   THE LEBESGUE INTEGRAL

At the turn of the twentieth century, Lebesgue was working on his doctoral dissertation while a professor at the Lycée Centrale, a girls' secondary school, in Nancy, France, from 1899 to 1901. Lebesgue had studied the works of Jordan, Borel, and Baire, and was building on their foundation; he also studied Riemann integration and Fourier series. The problem of certain derivatives not being integrable—functions failing the fundamental theorem—was Lebesgue's motivation. Dirichlet's characteristic function of the rationals (1829) gave an example that could not be Riemann integrated; but his function was nowhere continuous. Thomae's function (1875) is Riemann integrable and is continuous only on the irrationals. Riemann himself had given an example of an integrable function that was discontinuous at any rational whose denominator was divisible by 2 in his 1854 *Habilitation* thesis. Where is the demarcation between integrable and nonintegrable in terms of continuity? Lebesgue's theory answers these questions.

We'll start by classifying the functions to study, then define the Lebegue integral and investigate its properties.

Measurable Functions

Continuity is essentially a local condition describing the behavior of a function at a point. Uniform continuity extends the concept to sets, but still the focus is at the level of a point. We need a more global descriptor. In topology, continuity is described as the inverse image of an open set is an open set. We'll use a similar criterion to define a function as *measurable* in terms of measurable sets.

Theorem 3.15 (Measurability Condition for Functions) *Let f be an extended real-valued function that has a measurable domain D. The following are equivalent.*

1. *For each $r \in \mathbb{R}$, the set $\{x \in D \mid f(x) > r\} = f^{-1}((r, \infty))$ is measurable.*

2. *For each $r \in \mathbb{R}$, the set $\{x \in D \mid f(x) \geq r\} = f^{-1}([r, \infty))$ is measurable.*

3. *For each $r \in \mathbb{R}$, the set $\{x \in D \mid f(x) < r\} = f^{-1}((-\infty, r))$ is measurable.*

4. *For each $r \in \mathbb{R}$, the set $\{x \in D \mid f(x) \leq r\} = f^{-1}((-\infty, r]))$ is measurable.*

Proof: Let $D = \operatorname{dom}(f)$ be a measurable set.

Then $1 \Rightarrow 2$, since $\{x \mid f(x) \geq r\} = \bigcap_n \{x \mid f(x) > r - 1/n\}$, and the countable intersection of measurable sets is measurable.

Now $2 \Rightarrow 3$, because $\{x \mid f(x) < r\} = D - \{x \mid f(x) \geq r\}$, and the difference of two measurable sets is measurable.

Next, $3 \Rightarrow 4$, as $\{x \mid f(x) \leq r\} = \bigcap_n \{x \mid f(x) < r + 1/n\}$, and again, the countable intersection of measurable sets is measurable.

Last, $4 \Rightarrow 1$, since $\{x \mid f(x) > r\} = D - \{x \mid f(x) \leq r\}$, and, once more, the difference of two measurable sets is measurable. ∎

Figure 3.3 shows a function f and the set $A = \{x \mid f(x) > 2.5\}$; note the three components of A.

Corollary 3.16 *If f satisfies any of the measurability conditions, then for each $r \in \mathbb{R}$, the set $\{x \mid f(x) = r\}$ is measurable.*

The converse of the corollary is not true; even if the sets $\{x \mid f(x) = r\}$ are all measurable, the function need not satisfy the measurability conditions.

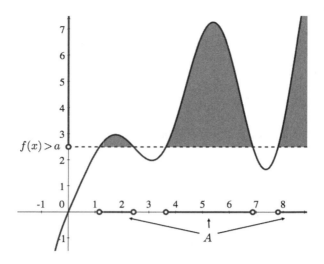

Figure 3.3   The Set $A = \{x \mid f(x) > 2.5\}$

Definition 3.5 (Lebesgue Measurable Function) *If an extended real-valued function f has a measurable domain and satisfies the measurability conditions, then f is called a* Lebesgue measurable *or a* measurable function.

■ **EXAMPLE 3.6   Step Functions**

A function $\phi : [a, b] \to \mathbb{R}$ is a *step function* if there is a partition $a = x_0 < x_1 < \cdots < x_n = b$ such that ϕ is constant on each interval $I_k = (x_{k-1}, x_k)$, then

$$\phi(x) = \sum_{k=1}^{n} a_k \chi_{I_k}(x)$$

where $a_k = \phi(I_k)$. See Figure 3.4. (We have not specified the values at x_k, a set of measure zero.) Since the set $\{x \mid f(x) \le r\}$ is a finite union of intervals, it is measurable. Thus, step functions are measurable.   ■

■ **EXAMPLE 3.7   Simple Functions**

A function $\psi : [a, b] \to \mathbb{R}$ is a *simple function* if the range of ψ is a finite set $\{a_1, a_2, \ldots, a_n\}$ and, for each $k = 1, \ldots, n$, $\psi^{-1}(a_k)$ is a measurable set. Since

$$\{x \mid \psi(x) > r\} = \bigcup_{a_k > r} \psi^{-1}(a_k)$$

and each $\psi^{-1}(a_k)$ is measurable, then ψ is a measurable function.

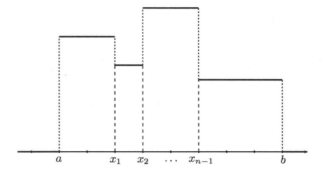

Figure 3.4 A Step Function

An alternate definition is ψ is a simple function if it can be written as

$$\psi(x) = a_1 \chi_{E_1}(x) + a_2 \chi_{E_2}(x) + \cdots + a_n \chi_{E_n}(x)$$

where $E_1, E_2, \ldots, E_n$ are pairwise-disjoint, measurable sets and $a_1, a_2, \ldots, a_n$ are constants.

1. The characteristic function of the rationals $\chi_{\mathbb{Q}}$ is a simple function, and thus measurable. *Would the characteristic function of a general measurable set be a simple function?*

2. Fix a value of $n \in \mathbb{N}$. Define $\psi_n : [0, n) \to \mathbb{R}$ for rational r by

$$\psi_n(r) = \sum_{k=1}^{n} k \cdot \chi_{[k-1,k)}(r)$$

and $\psi_n(x) = 0$ for irrational x. Then ψ_n is a simple function. *What does ψ_n's graph look like?*

■

Every step function is a simple function, but not conversely.

For a continuous function f with a measurable domain, the inverse image of an open set is open. Open sets are measurable. Hence, continuous functions with measurable domains are measurable functions.

Is the combination of measurable functions still measurable?

Theorem 3.17 (Algebra of Measurable Functions) *Let f and g be measurable functions on a common domain, and let $c \in \mathbb{R}$. Then $f + c$, cf, $f \pm g$, f^2, and fg are measurable.*

Proof: Since $\{x \mid f(x) + c < r\} = \{x \mid f(x) < r - c\}$ and, for nonzero c, $\{x \mid cf(x) < r\} = \{x \mid f(x) < r/c\}$, then $f + c$ and cf are measurable if and only if f is measurable. *What if $c = 0$?*

If $f(x) + g(x) < r$, then $f(x) < r - g(x)$. Therefore, there is rational number p so that $f(x) < p < r - g(x)$. *Why?* It follows that

$$\{x \mid f(x) + g(x) < r\} = \bigcup_{p \in \mathbb{Q}} \left(\{x \mid f(x) < p\} \cap \{x \mid g(x) < r - p\} \right)$$

which is a countable union of measurable sets, and hence measurable. Thus $f + g$ is measurable. The measurability of $f - g$ follows similarly.

If $r \geq 0$, then

$$\{x \mid f^2(x) > r\} = \{x \mid f(x) > \sqrt{r}\} \cup \{x \mid f(x) < -\sqrt{r}\}$$

and if $r < 0$, then

$$\{x \mid f^2(x) > r\} = \operatorname{dom}(f)$$

Hence, f^2 is measurable when f is.

The identity

$$fg = \frac{1}{4} \left((f + g)^2 - (f - g)^2 \right)$$

shows that fg is measurable when f and g are measurable.   ∎

Since integration is our goal, we also need to consider sequences and limits.

Theorem 3.18 *Suppose $\{f_n\}$ is a sequence of measurable functions with a common domain. Then $\sup\{f_1, f_2, \ldots f_n\}, \sup_n f_n, \inf\{f_1, f_2, \ldots f_n\}, \inf_n f_n, \limsup_n f_n,$ and $\liminf_n f_n$ are all measurable functions.*

Proof: Let $f = \sup\{f_1, f_2, \ldots f_n\}$. Then for any $r \in \mathbb{R}$,

$$\{x \mid f(x) > r\} = \bigcup_{k=1}^{n} \{x \mid f_k(x) > r\}$$

Explain why the set equality above is true! Since the finite union of measurable sets is measurable, f is a measurable function.

Let $g = \sup_n f_n$. Similarly, for any $r \in \mathbb{R}$,

$$\{x \mid g(x) > r\} = \bigcup_{k=1}^{\infty} \{x \mid f_k(x) > r\}$$

The countable union of measurable sets being measurable implies g is a measurable function.

Let $h = \limsup_n f_n$. Then h is measurable because

$$\limsup_n f_n = \inf_n \left(\sup_{k \geq n} f_k \right)$$

Proofs of the statements with infima are left to the exercises. ∎

Previously, we noted the characteristic function of $\mathbb{N}$ was zero almost everywhere, abbreviated as $\chi_\mathbb{N} = 0$ a.e., as an example of "equality almost everywhere." If two functions are equal almost everywhere and one is measurable, then so is the other.

Theorem 3.19 *Suppose f is a measurable function and $f = g$ a.e. Then g is a measurable function.*

Proof: Let $E = \{ x \mid f(x) \neq g(x) \}$. Then for $r \in \mathbb{R}$,

$$\{ x \mid g(x) > r \} = \{ x \mid f(x) > r \} \cup \{ x \in E \mid g(x) > r \} - \{ x \in E \mid g(x) \leq r \}$$

Explain why the set equality above holds! Both $\{ x \in E \mid g(x) > r \}$ and $\{ x \in E \mid g(x) \leq r \}$ are subsets of E, a set of measure zero. Therefore, both sets are measurable. Hence, if f is measurable, then so is g. ∎

Extend this concept to convergence.

Definition 3.6 (Convergence Almost Everywhere) *A sequence $\{ f_n \}$ converges to f almost everywhere if and only if the set of values for which $\{ f_n \}$ does not converge to f has measure zero. We write $f_n \to f$ a.e.*

■ **EXAMPLE 3.8**

Define $f_n : [0, 1] \to \mathbb{R}$ by

$$f_n(x) = \begin{cases} n & x \in \mathbb{Q} \\ \dfrac{1}{n} & \text{otherwise} \end{cases}$$

Then $f_n \to 0$ a.e. on $[0, 1]$. *Verify this!* ∎

Measurability is a very general property for a function; however, we will see that a measurable function cannot be too unstructured.

Theorem 3.20 *A function $f : [a, b] \to \mathbb{R}$ is measurable if and only if there is a sequence of simple functions $\{ \psi_n \}$ converging to f almost everywhere.*

Proof: Suppose f is measurable. Without loss of generality, assume that f is nonnegative. [Otherwise, write $f = f^+ - f^-$ where $f^+ = \max(f, 0)$ and $f^- = \max(-f, 0)$, and apply the argument below to each part.]

For $n \in \mathbb{N}$, define $A_{n,k}$ by

$$A_{n.k} = \left\{ x \, \middle| \, \frac{k-1}{2^n} \leq f(x) < \frac{k}{2^n} \right\}$$

for $k = 1, 2, \ldots, n2^n$, and

$$A_{0,n} = [a, b] - \bigcup_{k=1}^{2^n} A_{n,k}$$

Now let

$$\psi_n(x) = n \chi_{A_{0,n}}(x) + \sum_{k=1}^{n2^n} \frac{k-1}{2^n} \chi_{A_{n,k}}(x)$$

Figure 3.5 shows a function and its approximation by ψ_n. It's not difficult to see (*Verify these!*)

1. $\psi_1(x) \leq \psi_2(x) \leq \cdots$,

2. If $0 \leq f(x) \leq n$, then $|f(x) - \psi_n(x)| < 2^{-n}$,

3. $\lim_{n \to \infty} \psi_n(x) = f(x)$ a.e.

which proves this direction of the theorem.

Now suppose that a sequence of simple functions $\{\psi_n\}$ converges to f a.e. Since simple functions are measurable, then their limit g is measurable. Since $g = f$ a.e., then f is measurable. ∎

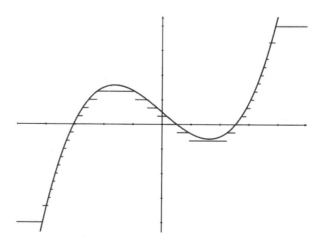

Figure 3.5   A Simple Approximation

We have actually shown that a nonnegative measurable function is the limit of a monotonically increasing sequence of simple functions. It can also be shown that a function is measurable if and only if it is the limit of step functions almost everywhere. Luzin's theorem (1912) showing measurable functions are nearly continuous is a direct consequence of the characterization of measurable functions in terms of step functions.

Theorem 3.21 (Luzin's Theorem) *Let f be a measurable function on* $[a, b]$. *For any* $\epsilon > 0$, *there is a continuous function g on* $[a, b]$ *such that*

$$\mu\left(\{x \mid f(x) \neq g(x)\}\right) < \epsilon$$

We'll defer proving Luzin's theorem as it requires the step function characterization and a theorem of Egorov that we'll see later.

We close our discussion of measurable functions with an interesting observation. A continuous function of a measurable function is measurable. However, a measurable function of a measurable function is not necessarily measurable. In fact, a measurable function of a continuous function need not be measurable.

Now that we've studied measurable functions, it's time to consider integration.

The Lebesgue Integral

Lebesgue's integral is based on partitioning the range, rather than the domain. In his 1926 address to the *Société Mathématique* in Copenhagen, Lebesgue said,

> Let us proceed according to the goal to be attained: to gather or group values of $f(x)$ which differ by little. It is clear then, that we must partition not (a, b), but rather the interval $(\underline{f}, \overline{f})$ bounded by the lower and upper bounds of f on (a, b).

[See Chae (1995, p. 234–248) for an English translation of Lebesgue's address.] Lebesgue used the diagram in Figure 3.6 to illustrate his idea. The range of f is partitioned. The inverse image of the y-axis interval consists of the four components appearing on the x-axis. If L is the length of the inverse image set, then the integral will be bounded by $\underline{f}_n \times L$ and $\overline{f}_n \times L$ where $\underline{f}_n$ and $\overline{f}_n$ are bounds of f on (y_n, y_n). Needing the length L led Lebesgue to develop measure.

Let's build an integral.

Integrals of Bounded Functions

There are properties of the Riemann integral that are very useful that we wish to keep when extending the definition. The Riemann integral of a characteristic function of an interval is the interval's length; i.e., $\int_a^b dx = b - a$. The counterpart of an interval is a measurable set. Also, the Riemann integral is linear: $\int_b^a (rf(x) + sg(x)) \, dx = r \int_b^a f(x) \, dx + s \int_a^b g(x) \, dx$. We build in these properties by basing the new integral on simple functions. To avoid dealing with infinities, we'll require the measure of the set $\{x \mid \psi(x) \neq 0\}$ to be finite; equivalently, ψ must be zero outside a set of finite measure. (Saying a set has finite measure implicitly assumes the set is measurable.) Recall that simple functions are measurable by definition.

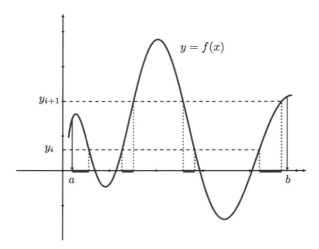

Figure 3.6   Lebesgue's Diagram

Definition 3.7 (Lebesgue Integral of a Bounded Function) *Let ψ be a simple function that is zero outside a set of finite measure, and let $\psi(x) = \sum_{k=1}^{n} a_k \chi_{A_k}(x)$ where each $A_k = \psi^{-1}(a_k)$ is the canonical representation of ψ. Define the* Lebesgue integral *of ψ by*

$$\int \psi d\mu \;=\; \sum_{k=1}^{n} a_k \mu(A_k)$$

If E is a measurable set, define the Lebesgue integral *of ψ over E by*

$$\int_{E} \psi d\mu \;=\; \int \psi \cdot \chi_E \, d\mu = \sum_{k=1}^{n} a_k \mu(A_k \cap E)$$

When the context is clear, we abbreviate the integrals as $\int \psi$ or $\int_E \psi$. Other notations that appear in the literature include $\int \psi(x) \, d\mu(x)$ and $\int \psi(x) \, d\mu$.

■ **EXAMPLE 3.9**

Dirichlet's "monster," the characteristic function of the rationals $\chi_{\mathbb{Q}}$ was the first non-Riemann integrable function on $[0, 1]$ we saw. Now, however, we have

$$\int_{[0,1]} \chi_{\mathbb{Q}} \, d\mu = \mu\left(\mathbb{Q} \cap [0, 1]\right) = 0$$

This function, even though discontinuous everywhere, is Lebesgue integrable! Further, if E is any measurable set, then $\mu(\mathbb{Q} \cap E) = 0$. Therefore we also have $\int_E \chi_{\mathbb{Q}} \, d\mu = 0$.   ■

Any simple function has multiple representations, so we need to demonstrate that all representations lead to the same value for the integral.

Theorem 3.22 *Let $\psi(x) = \sum_{k=1}^{n} a_k \chi_{E_k}(x)$ where each E_k has finite measure and the E_k are pairwise disjoint; that is, $E_k \cap E_j = \emptyset$ for $k \neq j$. Then*

$$\int \psi d\mu = \sum_{k=1}^{n} a_k \mu(E_k)$$

Proof: For any r, the set $\psi^{-1}(r) = \bigcup_{a_k = r} E_k$. Since Lebesgue measure is additive for disjoint sets, $\mu\left(\psi^{-1}(r)\right) = \sum_{a_k = r} \mu(E_k)$. Set $R = \text{range}(\psi)$. Then R is finite. *Why?* Therefore

$$\int \psi d\mu = \sum_{r \in R} r\mu\left(\psi^{-1}(r)\right)$$

$$= \sum_{r \in R} r \cdot \left(\sum_{a_k = r} \mu(E_k)\right)$$

$$= \sum_{k=1}^{n} a_k \mu(E_k)$$

Thus the integral of ψ is independent of representation.   ■

A development of the Lebesgue integral using partitions closely following the Riemann integral is given in Bear (1995).

Lebesgue integrals are linear and monotone by design.

Theorem 3.23 *Let ψ and ϕ be two simple functions that are both zero outside a set of finite measure, and let a and b be real constants. Then*

$$\int (a\psi + b\phi)d\mu = a\int \psi d\mu + b\int \phi d\mu$$

If $\psi(x) \leq \phi(x)$ a.e., then

$$\int \psi d\mu \leq \int \phi d\mu$$

Proof: We need to start by dealing with the simple function $a\psi + b\phi$. Let the canonical representation of ψ be $\psi(x) = \sum_{k=1}^{N_1} a_k \chi_{A_k}(x)$ and that of ϕ be $\phi(x) = \sum_{k=1}^{N_2} b_k \chi_{B_k}(x)$. Set $A_0 = \psi^{-1}(0)$ and $B_0 = \phi^{-1}(0)$. Form the set

$$\{E_k \mid k = 1, \ldots, N\} = \{A_i \cap B_j \mid i = 0, \ldots, N_1, j = 0, \ldots, N_2\}$$

Since $\{A_k\}$ and $\{B_k\}$ are pairwise disjoint sets, then $\{E_k\}$ is a pairwise disjoint collection. Thus,

$$\psi(x) = \sum_{k=1}^{N} a_k \chi_{E_k}(x) \qquad \text{and} \qquad \phi(x) = \sum_{k=1}^{N} b_k \chi_{E_k}(x)$$

Then we see that

$$a\psi(x) + b\phi(x) = \sum_{k=1}^{N}(aa_k + bb_k)\chi_{E_k}(x)$$

It now follows from the previous theorem that $\int (a\psi + b\phi)d\mu = a\int \psi d\mu + b\int \phi d\mu$.

Now, if $\psi(x) \leq \phi(x)$ a.e., then $\int \phi d\mu - \int \psi d\mu = \int(\phi - \psi)\, d\mu$ by part 1. Since $\phi(x) - \psi(x) \geq 0$ a.e., the definition of the integral gives $\int(\phi - \psi)\, d\mu \geq 0$, and part 2 is seen to hold.   ∎

Since the Lebesgue integral is linear, we don't need the requirement that the sets be disjoint in the representation of a simple function. *Why?*

In defining the Riemann integral, we used the maximum and minimum of the function f on the subinterval $[x_{k-1}, x_k]$ to build Riemann-Darboux sums. The upper Riemann integral is

$$\overline{\int_a^b} f(x)\, dx = \inf_P \sum_{k=1}^{n} \max_{[x_{k-1}, x_k]} f(x)\, \Delta x_k$$

We can look at this sum as the integral of a step function. *How?* Then

$$\overline{\int_a^b} f(x)\, dx = \inf_{\phi \geq f} \int_a^b \phi(x)\, dx$$

where ϕ is a step function. We use this reasoning, extend from step functions to simple functions, to build the Lebesgue integral of a function. For a bounded real-valued function f on a measurable set E, we check whether, for simple functions ψ and ϕ, the two expressions

$$\inf_{\psi \geq f} \int_E \psi d\mu \qquad \text{and} \qquad \sup_{\phi \leq f} \int_E \phi d\mu$$

are equal.

Theorem 3.24 *Suppose F is a bounded real-valued function on E, a set with finite measure. Then for all simple functions ψ and ϕ,*

$$\inf_{\psi \geq f} \int_E \psi d\mu = \sup_{\phi \leq f} \int_E \phi d\mu$$

if and only if f is measurable.

Proof: First, suppose that f is bounded by M and measurable. For $k = -n, -n + 1, -n + 2, \ldots, n - 1, n$, put

$$E_k = \left\{ x \; \middle| \; \frac{k-1}{n} \cdot M < f(x) \leq \frac{k}{n} \cdot M \right\}$$

The E_k are measurable (*Why?*), disjoint, and $E = \bigcup_k E_k$. Therefore

$$\sum_{k=-n}^{n} \mu(E_k) = \mu(E)$$

Set

$$\psi_n(x) = \sum_{k=-n}^{n} \frac{kM}{n} \chi_{E_k}(x) \quad \text{and} \quad \phi_n(x) = \sum_{k=-n}^{n} \frac{(k-1)M}{n} \chi_{E_k}(x)$$

The definition of ψ and ϕ implies that

$$\phi_n(x) \le f(x) \le \psi_n(x)$$

and $\psi_n(x) - \phi_n(x) \ge 0$. *Verify this!* Thence

$$\inf \int_E \psi d\mu \le \int_E \psi_n \, d\mu = \sum_{k=-n}^{n} \frac{kM}{n} \mu(E_k)$$

and

$$\sup \int_E \phi d\mu \ge \int_E \phi_n \, d\mu = \sum_{k=-n}^{n} \frac{(k-1)M}{n} \mu(E_k)$$

Subtracting the second inequality from the first and remembering $\phi_n - \psi_n \ge 0$ for all n, we arrive at

$$0 \le \inf \int_E \psi d\mu - \sup \int_E \phi d\mu \le \frac{M}{n} \sum_{k=-n}^{n} \mu(E_k) = \frac{M}{n} \mu(E)$$

Since n is arbitrary, this relation holds for all n. Hence

$$\inf \int_E \psi d\mu = \sup \int_E \phi d\mu$$

Now, for the other direction, we assume that $\inf \int_E \psi d\mu = \sup \int_E \phi d\mu$. For any n, there are simple functions ψ_n and ϕ_n such that $\phi_n(x) \le f(x) \le \psi_n(x)$ and

$$\int \psi_n \, d\mu - \int \phi_n \, d\mu < \frac{1}{n}$$

Why? The functions

$$\hat{\psi} = \inf \psi_n \quad \text{and} \quad \hat{\phi} = \sup \phi_n$$

are measurable and $\hat{\phi}(x) \le f(x) \le \hat{\psi}(x)$. Let $F = \{x \,|\, \hat{\phi}(x) < \hat{\psi}(x)\}$ and $F_r = \{x \,|\, \hat{\phi}(x) < \hat{\psi}(x) - 1/r\}$. It follows that $F = \bigcup_r F_r$. Since

$$\phi_n(x) \le \hat{\phi}(x) \le f(x) \le \hat{\psi}(x) \le \psi_n(x)$$

we have $F_r \subseteq \{x \mid \phi_n(x) < \psi_n(x) - 1/r\}$ which has measure less than r/n. Why? Then $\mu(F_r) = 0$ as n is arbitrary. Hence $\mu(F) = 0$, and $\hat{\phi} = \hat{\psi}$ a.e. Therefore $\hat{\phi} = f$ a.e., which implies that f is measurable. ∎

So, if a function f is measurable and bounded, then the analogues of the upper and lower integrals are equal. We can use this fact to define the integral of f.

Definition 3.8 *If f is a bounded measurable function defined on a set E having finite measure, then the* Lebesgue integral *of f over E is*

$$\int_E f \, d\mu = \inf \int_E \psi d\mu$$

for all simple functions $\psi \geq f$.

We already know that Lebesgue's integral can handle functions that Riemann's cannot. Is every Riemann integrable function also Lebesgue integrable? Recall we interpreted Riemann sums as integrals of step functions. The answer now comes easily.

Theorem 3.25 *Let $f : [a, b] \to \mathbb{R}$ be bounded and Riemann integrable on $[a, b]$. Then f is measurable and*

$$\int_a^b f(x) \, dx = \int_{[a,b]} f \, d\mu$$

Proof: Every step function is a simple function. Interpret Riemann sums as step functions to yield

$$\underline{\int}_a^b f(x) \, dx \leq \sup_{\phi \leq f} \int_{[a,b]} \phi d\mu \leq \inf_{\psi \geq f} \int_{[a,b]} \psi d\mu \leq \overline{\int}_a^b f(x) \, dx$$

The inequalities are actually equalities given the Riemann integrability of f. Then f is measurable by Theorem 3.24. ∎

This result tells us that Lebesgue's integration is an extension of Riemann's: we have added to the class of integrable functions. Since the Lebesgue and Riemann integrals are equal, we normally write $\int_a^b f \, d\mu$ rather than $\int_{[a,b]} f \, d\mu$.

Let's collect several properties that follow immediately from the definition.

Theorem 3.26 (Properties of the Lebesgue Integrals of Bounded Functions) *Let f and g be bounded measurable functions both defined on a set E of finite measure, and let $c \in \mathbb{R}$. Then*

1. $\displaystyle\int_E cf \, d\mu = c \int_E f \, d\mu.$

2. $\displaystyle\int_E f + g \, d\mu = \int_E f \, d\mu + \int_E g \, d\mu.$

3. *If $f \leq g$ a.e., then $\int_E f \, d\mu \leq \int_E g \, d\mu$.*

4. *If $f = g$ a.e., then $\int_E f \, d\mu = \int_E g \, d\mu$.*

5. *If $m \leq f(x) \leq M$, then*

$$m \cdot \mu(E) \leq \int_E f \, d\mu \leq M \cdot \mu(E)$$

6. *If E_1 and E_2 are disjoint measurable sets with $E_1 \cup E_2 = E$, then*

$$\int_{E_1} f \, d\mu + \int_{E_2} f \, d\mu = \int_E f \, d\mu$$

The proofs are straightforward and are in Exercise 3.22.

Corollary 3.27 *If f is a bounded measurable function defined on a set E of finite measure, then*

$$\left| \int_E f \, d\mu \right| \leq \int_E |f| \, d\mu$$

We end this segment with a theorem that characterizes Riemann integration by using Lebesgue's theory. Riemann's condition for integrability is based on the sets on which a function varies more than a small amount. If the total length of these sets is small enough, the function is integrable. The *oscillation of a function f on the interval I* is given by

$$\omega(f, I) = \sup_{s,t \in I} |f(s) - f(t)|$$

and the *oscillation of f at the point x* is

$$\omega(f, x) = \inf_{\delta > 0} \omega(f, N_\delta(x))$$

Recall $N_\delta(x) = (x - \delta, x + \delta)$. Riemann's theorem is stated here without proof for comparison to Lebesgue's condition.

Theorem 3.28 (Riemann's Criterion for Riemann Integrability) *Let the function f be bounded on the interval $[a, b]$. Then f is Riemann integrable if and only if for any $\epsilon, \sigma > 0$ there is a $\delta > 0$ such that for any partition P with $\|P\| < \delta$ it follows that*

$$\sum_{\{j \,|\, \omega(f, I_j) > \sigma\}} \Delta x_j < \epsilon$$

where $I_j = [x_{j-1}, x_j]$.

Riemann's criterion in words is: f is integrable if and only if the sum of the lengths of the intervals where f varies more than ϵ is less than σ whenever the partition's norm is less than δ. That is, with tongue firmly in cheek, f is integrable if and only if it misbehaves in only a small collection of tiny neighborhoods.

Lebesgue recognized that Riemann's condition was really characterizing continuity. A function needed to be "almost continuous" to be Riemann integrable.

Theorem 3.29 (Lebesgue's Criterion for Riemann Integrability) *Let the function f be bounded on the interval $[a, b]$. Then f is Riemann integrable if and only if f is continuous almost everywhere on $[a, b]$.*

Proof: First, suppose f is Riemann integrable and let $\epsilon > 0$. Define the set N_r by

$$N_r = \{x \mid \omega(f, x) > r\}$$

Then any interval I that contains a point of N_r will have $\omega(f, I) \geq r$. Since f is Riemann integrable, there is a partition P so that the difference between the upper and lower Riemann sums is less than $r\epsilon/2$. Set $I_j = [x_{j-1}, x_j]$. Then

$$\sum_{\{j \mid I_j \cap N_r \neq \emptyset\}} \omega(f, I_j)\Delta x_j \leq \sum_{k=1}^{n} \omega(f, I_j)\Delta x_j < \frac{r\epsilon}{2}$$

Now, since the sum is less than $r\epsilon/2$, and $\omega(f, I_j) > r$ (*Why?*), we must have

$$\sum_{\{j \mid I_j \cap N_r \neq \emptyset\}} \Delta x_j < \frac{r\epsilon}{2r} = \frac{\epsilon}{2}$$

Each point of N_r is in some $I_j = [x_{j-1}, x_j]$; we can cover N_r with the collection of open sets $\{(x_{j-1}, x_j) \mid (x_{j-1}, x_j) \cap N_r \neq \emptyset\} \cup \{ (x_j - \epsilon/(2n), x_j + \epsilon/(2n))\}$ which has a total length less than ϵ. Since $\epsilon > 0$ is arbitrary, N_r must have measure zero. The set N of points of discontinuity of f is the countable union $\bigcup_{k=1}^{\infty} N_{1/k}$. Therefore N has measure zero.

For the other direction, suppose f is continuous almost everywhere on $[a, b]$ and $|f(x)| \leq M$. Let $\epsilon > 0$ be given. This part of the proof uses Riemann's idea of breaking the sum into two parts, one with small oscillations, the other with small intervals. Set $r = \epsilon/(2(b - a))$, and now let

$$N_r = \{x \mid \omega(f, x) \geq r\}$$

Thus N_r is a closed set of measure zero. *Why?* Hence we can cover N_r with a finite set $\mathcal{F}$ of disjoint closed intervals of total length less than $\epsilon/(4M + 1)$. Any point x of $[a, b]$ not covered by these intervals has oscillation less than $r = \epsilon/(2(b - a))$. Choose x_k from the remaining points to add to the endpoints of the intervals in $\mathcal{F}$ to form a partition of $[a, b]$. Intervals of the partition are either in $\mathcal{F}$ or have

$(\sup f - \inf f) \leq \epsilon/(2(b-a))$. Then

$$\sum_{k=1}^{n} \left(\sup_{I_k} f(x) - \inf_{I_k} f(x) \right) \Delta x_k \leq \sum_{\{k \,|\, I_k \notin \mathcal{F}\}} \frac{\epsilon}{2(b-a)} \Delta x_k + \sum_{\{k \,|\, I_k \in \mathcal{F}\}} 2M \Delta x_k$$

$$\leq \frac{\epsilon}{2(b-a)} \sum_{k=1}^{n} \Delta x_k + 2M \sum_{\{k \,|\, I_k \in \mathcal{F}\}} \Delta x_k$$

$$< \frac{\epsilon}{2} + 2M \cdot \frac{\epsilon}{4M+1} < \epsilon$$

Then, by Theorem 2.36, f is Riemann integrable.   ∎

With the results for bounded functions in hand, we can extend integration to unbounded functions, the analogue of improper Riemann integrals.

Integrals of Unbounded Functions   First, we define Lebesgue integrals of nonnegative measurable functions that may be unbounded; then we deal with general measurable functions by splitting the function into positive and negative parts, working with each separately. A function may be unbounded in different ways: a function may have an unbounded range or an unbounded domain. We focus on functions with an unbounded range; analogous techniques may be applied to handle unbounded domains.

Define the *n-truncation* of a nonnegative function f on the interval $[a,b]$ to be the minimum of $f(x)$ and n. Clearly, the sequence $\{f_n\}$ of n-truncations converges monotonically to f on $[a,b]$. Now $f_n = \min\{f, n\}$ is measurable and bounded for each n. *Why?* Hence $\int_a^b f_n \, d\mu$ exists for all n.

Definition 3.9 (Lebesgue Integral of an Unbounded Function) *Let f be a nonnegative measurable function on $[a,b]$, and let $\{f_n\}$ be the sequence of n-truncations of f. Then*

$$\int_a^b f \, d\mu = \lim_{n \to \infty} \int_a^b f_n \, d\mu$$

which may be infinite. If the limit is finite, we say f is Lebesgue integrable on $[a,b]$.

■ **EXAMPLE 3.10**

Determine $\int_0^2 f \, d\mu$ for

$$f(x) = \begin{cases} \dfrac{1}{(x-1)^{2/3}} & x \neq 1 \\ +\infty & x = 1 \end{cases}$$

The n-truncations of f are

$$
f_n(x) = \begin{cases}
\dfrac{1}{(x-1)^2} & 0 \le x < 1 - \dfrac{1}{n^{3/2}} \\[3mm]
n & 1 - \dfrac{1}{n^{3/2}} \le x \le 1 + \dfrac{1}{n^{3/2}} \\[3mm]
\dfrac{1}{(x-1)^2} & 1 + \dfrac{1}{n^{3/2}} < x \le 2
\end{cases}
$$

Graph f and several f_n. Then f_n is continuous, and hence Riemann integrable. Thus

$$
\int_0^2 f_n \, d\mu = \int_0^{1-1/n^{3/2}} f_n \, d\mu + \int_{1-1/n^{3/2}}^{1+1/n^{3/2}} n \, d\mu + \int_{1+1/\sqrt{n^3}}^2 f_n \, d\mu
$$

$$
= \left(3 - \frac{3}{\sqrt{n}}\right) + \frac{2}{\sqrt{n}} + \left(3 - \frac{3}{\sqrt{n}}\right)
$$

$$
= 6 - \frac{4}{\sqrt{n}}
$$

Letting n go to infinity gives us $\displaystyle\int_0^2 f \, d\mu = 6$.   ∎

Does the improper Riemann integral of f from Example 3.10 *converge?*

Does the algebra of integrals still apply to unbounded nonnegative functions? Yes!

Theorem 3.30 *If f and g are nonnegative measurable functions, and $c > 0$, then*

1. $\displaystyle\int_a^b cf \, d\mu = c \int_a^b f \, d\mu$,

2. $\displaystyle\int_a^b f + g \, d\mu = \int_a^b f \, d\mu + \int_a^b g \, d\mu$.

3. *If $f \le g$ a.e., then $\displaystyle\int_a^b f \, d\mu \le \int_a^b g \, d\mu$.*

The proofs are straightforward applications of Theorem 3.26 and are left to the exercises.

For functions that are both positive and negative we use the device from the proof of Theorem 3.20 of splitting the function into positive and negative parts. Define the *positive part of f* to be $f^+(x) = \max\{f(x), 0\}$ and the *negative part of f* to be $f^-(x) = \max\{-f(x), 0\}$. Both f^+ and f^- will be nonnegative functions.

Definition 3.10 *Let* f *be a measurable function on* $[a, b]$. *Then*

$$\int_a^b f \, d\mu = \int_a^b f^+ \, d\mu - \int_a^b f^- \, d\mu$$

which may be infinite. If the integrals of f^+ *and* f^- *are both finite, so is the integral of* f. *Then we say* f *is* Lebesgue integrable *on* $[a, b]$, *and write* $f \in \mathcal{L}([a, b])$. *If both* $\int_a^b f^+ \, d\mu$ *and* $\int_a^b f^- \, d\mu$ *are infinite, we say the integral of* f *does not exist.*

Combine the two facts

$$f \in \mathcal{L}([a, b]) \iff f^+ \text{ and } f^- \in \mathcal{L}([a, b])$$
$$|f| = f^+ + f^-$$

to see that f is Lebesgue integrable if and only if $|f|$ is. Is this relation also true for Riemann inegrals? Check what happens with the function $R(x) = 1$ if $x \in \mathbb{Q}$ and $R(x) = -1$ otherwise.

The standard way to integrate f over a measurable subset of $[a, b]$ is to multiply f by the characteristic function of the set.

Definition 3.11 *If* f *is a measurable function and* E *is a measurable subset of* $[a, b]$, *then*

$$\int_E f \, d\mu = \int_a^b f \cdot \chi_E \, d\mu$$

Observe that a set of measure zero will not affect the value of a Lebesgue integral since

$$\mu(E) = 0 \implies \int_E f \, d\mu = 0$$

Verify this! Dirichlet's and Thomae's functions both have Lebesgue integral equal to zero over $[0, 1]$ (or over any finite interval for that matter). Compare this result to Riemann integrals of each: Dirichlet's function is *not* Riemann integrable, while Thomae's is! *Why doesn't this contradict Theorem 3.29?*

The expected theorem on the algebra of general Lebesgue integrals is left to the reader as an exercise.

We turn to studying the advantages for convergence gained by Lebesgue's integral.

3.3 MEASURE, INTEGRAL, AND CONVERGENCE

One of the main motivations in the development of integration was the question of convergence of Fourier series. The expression

$$S_n(x) = \frac{1}{2\pi} \int_{-\pi}^{\pi} f(t) \frac{\sin \left((n + \frac{1}{2})(x - t) \right)}{\sin \left(\frac{1}{2}(x - t) \right)} \, dt$$

is Dirichlet's integral form of the nth partial sum of f's Fourier series. Letting f_n be the integrand brings us to question what properties does f need to have to guarantee convergence for S_n. We are led to investigating the relation between integration and sequences of functions.

To learn more about how Fourier series influenced the development of real analysis, see Bressoud (2005, 2008). A very readable exposition of Fourier series appears in Jackson (1941); also look ahead at Section 4.3 for an introduction.

Types of Convergence

There are several different modes of convergence that we are led to consider. In beginning real analysis, we focused on pointwise and uniform convergence. We have seen convergence almost everywhere earlier in this chapter. Now we define new types by changing how we measure the "distance" between functions.

Definition 3.12 (Types of Convergence) *Let* $\{f_n\}$ *be a sequence of functions. Then*

Almost Everywhere: f_n *converges to* f almost everywhere *if and only if* $f_n(x) \to f(x)$ *except on a set of measure zero.*

Almost Uniformly: f_n *converges to* f almost uniformly *if and only if for every* $\epsilon > 0$ *there is a set* E_ϵ *of measure less than* ϵ *such that* $f_n \to f$ *uniformly on* E_ϵ^c, *the complement of* E_ϵ.

In Mean: f_n *converges to* f in mean *if and only if*

$$\lim_{n \to \infty} \int |f_n - f| \, d\mu = 0$$

In Measure: f_n *converges to* f in measure *if and only if for every* $\epsilon > 0$

$$\lim_{n \to \infty} \mu \{x \mid |f_n(x) - f(x)| > \epsilon\} = 0$$

Figure 3.7 summarizes the relation between the different modes of convergence. Each arrow in the diagram represents a convergence theorem. For instance, almost uniform convergence implies both convergence almost everywhere and convergence in measure. If we restrict sequences to bounded domains like $[a, b]$, several more arrows could be drawn. Soon we'll add an arrow with a 1911 theorem of Egorov's. See Bressoud (2008) or Royden (1988) for more details and further theory.

■ **EXAMPLE 3.11**

1. Define the sequence $\{f_n\}$ by $f_1 = \chi_{[0,1]}$, $f_2 = \chi_{[0,1/2]}$, $f_3 = \chi_{[1/2,1]}$, $f_4 = \chi_{[0,1/3]}$, $f_5 = \chi_{[1/3,2/3]}$, $f_6 = \chi_{[2/3,1]}$, and so forth. Figure 3.8 shows graphs of f_1 through f_6. Then, following the order diagrammed in Figure 3.7,

   (a) f_n does not converge uniformly on $[0, 1]$.

   (b) $f_n \to 0$ in mean on $[0, 1]$.

   (c) f_n does not converge almost uniformly on $[0, 1]$.

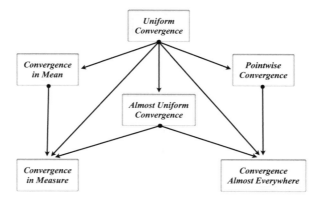

Figure 3.7   Types of Convergence

(d) f_n does not converge pointwise on $[0, 1]$ or even at any $x \in [0, 1]$.

(e) $f_n \to 0$ in measure on $[0, 1]$.

(f) f_n does not converge almost everywhere on $[0, 1]$.

Verify each claim above!

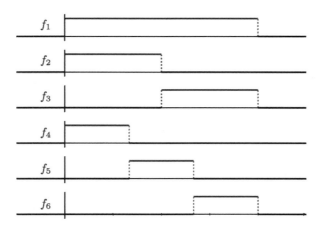

Figure 3.8   Convergence in Mean and Measure

2. Let

$$g_n(x) = \frac{n^2 x}{1 + n^3 x^2}$$

Graphs of $g_n(x)$ are in Figure 3.9. Then

(a) g_n does not converge uniformly on $[0, 1]$.

(b) $g_n \to 0$ in mean on $[0, 1]$.

(c) $g_n \to 0$ almost uniformly on $[0, 1]$. *Show that* $\{g_n\}$ *converges uniformly on* $[\epsilon, 1] = [0, 1] - N_\epsilon(0)$ *for any* $\epsilon > 0$.

(d) $g_n \to 0$ pointwise on $[0, 1]$.

(e) $g_n \to 0$ in measure on $[0, 1]$.

(f) $g_n \to 0$ almost everywhere on $[0, 1]$.

Verify each claim above!   ■

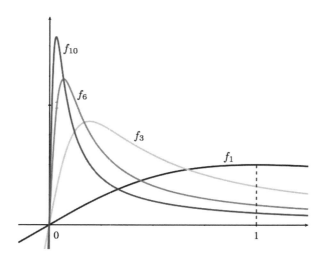

Figure 3.9   Almost Uniform, But Not Uniform Convergence

There is a general discussion of different types of convergence and their relationships in Bartle (1995, Chapter 7).

The first relation we would add to the diagram is: converging pointwise is nearly converging uniformly for a sequence of functions defined on a bounded interval $[a, b]$.

Theorem 3.31 (Egorov's Theorem) *If the sequence of measurable functions* $\{f_n\}$ *converges to* f *almost everywhere on* $[a, b]$, *then it converges almost uniformly to* f *on* $[a, b]$; *that is, for any* $\epsilon > 0$, *there is a set* $E \subset [a, b]$ *with* $\mu(E) < \epsilon$ *such that* f_n *converges to* f *uniformly on* $[a, b] - E$.

Proof: Following Bressoud (2008, p. 193), we do two reductions to simplify the proof. First, $f_n \to f$ a.e. on $[a, b]$ if and only if $f_n - f \to 0$ a.e. on $[a, b]$. Thus, we can assume $f = 0$ without loss of generality. Second, define $g_n(x) = \sup_{m \geq n} |f_m(x)|$. Then $g_n \to 0$ uniformly on E if and only if $f_n \to 0$ uniformly on E. *Verify this!* Since $\{g_n\}$ is made of supremums of smaller and smaller sets, they must monotonically decrease to zero. Let A be the set of measure zero where $f_n \not\to 0$; therefore $g_n \not\to 0$ on A.

Define

$$E_{k,n} = \left\{ x \,|\, 0 \leq g_n(x) < \frac{1}{2^k} \right\}$$

Then each $E_{k,n}$ is measurable, and

$$E_{k,1} \subseteq E_{k,2} \subseteq E_{k,3} \subseteq \cdots$$

because g_n monotonically decreases to zero. Further, since $g_n \to 0$ on $[a, b] - A$, we have that

$$[a, b] - A \subseteq \bigcup_{n=1}^{\infty} E_{k,n}$$

Since $\mu(A) = 0$, and the limit of the measures of nested sets is the measure of the limit of the sets (*By which theorem?*),

$$b - a \leq \mu \left(\bigcup_{n=1}^{\infty} E_{k,n} \right) = \lim_{n \to \infty} \mu(E_{k,n})$$

Let $\epsilon > 0$ be given. For each $k \in \mathbb{N}$, there is an $n = n_k$ such that

$$\mu(E_{k,n}) > (b - a) - \frac{\epsilon}{2^k}$$

Why? Now define

$$E = [a, b] - \bigcap_{k=1}^{\infty} E_{k,n_k} = \bigcup_{k=1}^{\infty} ([a, b] - E_{k,n_k})$$

Then $\mu(E)$ is bounded by

$$\mu(E) \leq \mu \left(\bigcup_{k=1}^{\infty} ([a, b] - E_{k,n_k}) \right) \leq \sum_{k=1}^{\infty} \frac{\epsilon}{2^k} = \epsilon$$

i.e., the measure of E is less than ϵ. For any $x \in [a, b] - E$, we have $x \in \bigcap_{k=1}^{\infty} E_{k,n_k}$. Hence $x \in E_{k,n_k}$ for every k. Therefore, for every $m > n_k$,

$$0 \leq g_m(x) \leq g_{n_k}(x) < \frac{1}{2^k}$$

Thus g_n converges uniformly to zero on $[a, b] - S$, and the theorem is proved. ∎

Lebesgue Integration and Convergence

Lebesgue's integral expands the class of functions that are integrable. Does adding more functions to the class change the requirements for interchanging integrals and limits? We'll see the answer gives an enhanced theory of convergence.

If each function in a sequence is bounded by the same value, we call the sequence *uniformly bounded*. Our first convergence theorem is for uniformly bounded sequences.

Theorem 3.32 (Bounded Convergence Theorem) *If $\{f_n\}$ is a uniformly bounded sequence of measurable functions converging to f a.e. on $[a, b]$, then f is measurable and*

$$\lim_{n \to \infty} \int_a^b f_n \, d\mu = \int_a^b \lim_{n \to \infty} f_n \, d\mu = \int_a^b f \, d\mu$$

Proof: On the set of measure zero where $f_n \not\to f$, define f to be zero. Let M be a bound for f_n on $[a, b]$. Since f is the limit of bounded measurable functions, it must be bounded and measurable. Therefore, f is integrable.

Let $\epsilon > 0$. Egorov's theorem implies there is a set E with $\mu(E) < \epsilon$ so that $f_n \to f$ uniformly on $[a - b] - E$. Then, for n large enough,

$$\left| \int_a^b f_n \, d\mu - \int_a^b f \, d\mu \right| \leq \int_a^b |f_n - f| \, d\mu$$

$$= \int_E |f_n - f| \, d\mu + \int_{[a,b]-E} |f_n - f| \, d\mu$$

$$\leq 2M\mu(E) + \epsilon\,\mu\,([a,b] - E)$$

$$= [2M + \mu\,([a,b] - E)]\,\epsilon < K\epsilon$$

Since $\epsilon > 0$ is arbitrary, the integrals must be equal. ∎

If instead of bounding the functions by a constant, we ask that the functions be nonnegative and bound each by the next in the sequence, we can obtain the same result. This requirement is simply that the sequence is monotone increasing and nonnegative. Beppo Levi proved this theorem in 1906. First, we prove a lemma due to Fatou (1906) that is very useful in itself.

Lemma 3.33 (Fatou's Lemma) *If $\{f_n\}$ is a sequence of nonnegative measurable functions converging to f a.e. on $[a, b]$, then*

$$\int_a^b f \, d\mu = \int_a^b \lim_{n \to \infty} f_n \, d\mu \leq \liminf_{n \to \infty} \int_a^b f_n \, d\mu$$

Proof: Without loss of generality, assume $f_n \to f$ for all x in $[a, b]$ as sets of measure zero do not affect integrals. Let ψ be a measurable function on $[a, b]$ that is less than or equal to f and is bounded by, say, M. Set

$$\psi_n(x) = \min\{f_n(x), \psi(x)\} \leq f_n(x)$$

Then $\{\psi_n\}$ is a uniformly bounded sequence of measurable functions. By the bounded convergence theorem,

$$\int_a^b \lim_{n\to\infty} \psi_n \, d\mu = \lim_{n\to\infty} \int_a^b \psi_n \, d\mu \leq \liminf_{n\to\infty} \int_a^b f_n \, d\mu$$

Take the supremum over all bounded measurable functions ψ such that $\psi \leq f$ to obtain

$$\int_a^b f \, d\mu \leq \liminf_{n\to\infty} \int_a^b f_n \, d\mu$$

∎

Levi's theorem is also called the *monotone convergence theorem.*

Theorem 3.34 (Levi's Theorem) *If $\{f_n\}$ is a monotone increasing sequence of nonnegative measurable functions converging to f a.e. on $[a, b]$, then f is measurable and*

$$\lim_{n\to\infty} \int_a^b f_n \, d\mu = \int_a^b \lim_{n\to\infty} f_n \, d\mu = \int_a^b f \, d\mu$$

Proof: Fatou's lemma gives

$$\int_a^b f \, d\mu \leq \liminf_{n\to\infty} \int_a^b f_n \, d\mu$$

On the other hand, for each n, we have $f_n \leq f$. Therefore $\int_a^b f_n \, d\mu \leq \int_a^b f \, d\mu$. Take the supremum of both sides to see

$$\limsup_{n\to\infty} \int_a^b f_n \, d\mu \leq \int_a^b f \, d\mu$$

Thus

$$\int_a^b f \, d\mu = \lim_{n\to\infty} \int_a^b f_n \, d\mu$$

∎

A clever proof of Levi's theorem using double sequences of simple functions rather than Fatou's lemma appears in Bressoud (2008, p. 174) and Burk (2007, p. 122).

■ **EXAMPLE 3.12**

Find $\int_0^1 x \ln\left(1/x^4\right) d\mu$ using

$$f_n(x) = \begin{cases} x \ln\left(\dfrac{1}{x^4}\right) & \dfrac{1}{n} \leq x \leq 1 \\ 0 & 0 \leq x < \dfrac{1}{n} \end{cases}$$

Since f_n is nonnegative and monotone increasing (*Show this! Also, graph several f_n's.*), we can apply Levi's theorem. Now

$$\int_0^1 f_n \, d\mu = \int_{1/n}^1 f_n \, d\mu$$

$$= \int_{1/n}^1 x \ln\left(\frac{1}{x^4}\right) \, dx$$

where the last integral is a Riemann integral. *Why is this change valid?* Evaluating, we see

$$\int_0^1 f_n \, d\mu = 1 - \left(\frac{1}{n^2} + \frac{2\ln(n)}{n^2}\right)$$

which goes to 1 as $n \to \infty$. Hence, by Levi's theorem, the original integral is equal to 1. ∎

A nice corollary comes from rephrasing Levi's theorem in terms of series.

Corollary 3.35 *Suppose $\{f_n\}$ is a sequence of nonnegative measurable functions on $[a, b]$. Then*

$$\int_a^b \sum_{k=1}^\infty f_n(x) \, d\mu = \sum_{k=1}^\infty \int_a^b f_n(x) \, d\mu$$

Riemann integrals required uniform convergence in order to exchange of summation and integration.

Our last result on integration and convergence is the very powerful *Lebesgue's dominated convergence theorem* of 1910. Lebesgue was able to replace "monotone and bounded" with "bounded by an integrable function."

Theorem 3.36 (Lebesgue's Dominated Convergence Theorem) *Let $\{f_n\}$ be a sequence of integrable functions converging to f a.e. on $[a, b]$. If there is an integrable function g on $[a, b]$ such that $|f_n| \le g$ for all n, then f is integrable on $[a, b]$, and*

$$\int_a^b f \, d\mu = \int_a^b \lim_{n\to\infty} f_n \, d\mu = \lim_{n\to\infty} \int_a^b f_n \, d\mu$$

Proof: Since f_n is integrable, it is measurable. Therefore f, the limit of the f_n's, is measurable and integrable. As before, we can adjust f on a set of measure zero without loss of generality; thus, we can assume f is finite and is the limit of f_n on $[a, b]$.

Define two monotone sequences, one increasing and one decreasing, by

$$\underline{f}_n = \inf\{f_n, f_{n+1}, \dots\} \quad \text{and} \quad \overline{f}_n = \sup\{f_n, f_{n+1}, \dots\}$$

Then $-g \le \underline{f}_n \le f_n \le \overline{f}_n \le g$. So both $\underline{f}_n$ and $\overline{f}_n$ are integrable. Now, both sequences $\{g + \underline{f}_n\}$ and $\{g - \overline{f}_n\}$ are nonnegative and monotone increasing. *Why?*

Therefore both converge to integrable functions, namely $g + \underline{f}_n \to g + f$ and $g - \overline{f}_n \to g - f$.

Apply Levi's theorem to see

$$\int_a^b (g + f)\, d\mu = \lim_{n \to \infty} \int_a^b (g + \underline{f}_n)\, d\mu$$

$$= \int_a^b g\, d\mu + \int_a^b \lim_{n \to \infty} \underline{f}_n\, d\mu$$

Subtract the $\int_a^b g\, d\mu$ from both sides to have

$$\int_a^b f\, d\mu = \int_a^b \lim_{n \to \infty} \underline{f}_n\, d\mu$$

Similarly,

$$\int_a^b f\, d\mu = \int_a^b \lim_{n \to \infty} \overline{f}_n\, d\mu$$

On the other hand, $\underline{f}_n \leq f_n \leq \overline{f}_n$ implies that

$$\int_a^b \underline{f}_n\, d\mu \leq \int_a^b f_n\, d\mu \leq \int_a^b \overline{f}_n\, d\mu$$

for all n. Hence $\int_a^b f\, d\mu = \lim_{n \to \infty} \int_a^b f_n\, d\mu$, and the theorem holds. ∎

■ **EXAMPLE 3.13**

Show that

$$\lim_{n \to \infty} \int_0^1 \frac{nx}{1 + n^2 x^2}\, d\mu = 0$$

Let $f_n(x) = nx/(1 + n^2 x^2)$, First, we look at a graph to gain insight. Figure 3.10 shows f_n for $n = 1, 3, 6, 10$. The graphs suggest f_n is bounded by $g(x) = 1/2$. A little elementary calculus verifies this bound.

$$f_n'(x) = \frac{n\left(1 - n^2 x^2\right)}{\left(1 + n^2 x^2\right)^2}$$

So f_n' is zero when $x = 1/n$. Then $f_n(1/n) = 1/2$ shows domination by $g(x) = 1/2$. *Does the bounded convergence theorem also apply?*

By Lebesgue's dominated convergence theorem,

$$\lim_{n \to \infty} \int_0^1 \frac{nx}{1 + n^2 x^2}\, d\mu = \int_0^1 \left(\lim_{n \to \infty} \frac{nx}{1 + n^2 x^2} \right) d\mu$$

$$= \int_0^1 0\, d\mu = 0$$

Verify that $\lim_{n \to \infty} nx/(1 + n^2 x^2) = 0$ *for all* $x \in [0, 1]$! ∎

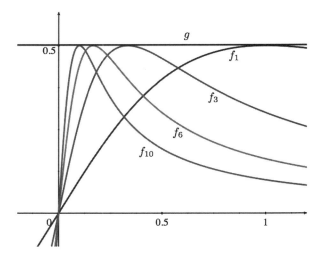

Figure 3.10 Plot of $f_n(x) = nx/(1 + n^2x^2)$ for Several n

Let's also express Lebesgue's dominated convergence theorem in terms of series.

Corollary 3.37 *Let* $\{f_n\}$ *be a sequence of integrable functions such that* $\sum_n f_n$ *converges a.e. on* $[a, b]$. *If there is an integrable function* g *on* $[a, b]$ *such that*

$$\left| \sum_{k=1}^{N} f_k(x) \right| \leq g(x) \ a.e.$$

for all N, *then series* $\sum_n f_n$ *is integrable on* $[a, b]$, *and*

$$\int_a^b \left(\sum_{k=1}^{\infty} f_k \right) d\mu = \sum_{k=1}^{\infty} \left(\int_a^b f_k \, d\mu \right)$$

Lebesgue has shown that we can replace uniform convergence with measurability and a dominating function yet still interchange limits and integrals. This very powerful theorem helped to answer many questions in diverse areas of analysis.

We end our study of Lebesgue's theory by stating the last theorem from Lebesgue's text *Leçons sur l'intégration* that uses the power of Lebesgue integration and convergence in a result for elementary calculus.

Theorem 3.38 *A monotonic function has a derivative almost everywhere.*

Look to Boas (1981, p. 155) for a proof.

In the next section, we ask how strange are these new concepts of measure and convergence really?

3.4 LITTLEWOOD'S THREE PRINCIPLES

John E. Littlewood (1885–1977) was a well-respected, highly prolific, classical analyst, which means his research was predominantly in real and complex analysis, as opposed to functional analysis; however, he also worked in other areas such as analytic number theory and differential equations. In 1944, Littlewood wrote a very influential textbook *Lectures on the Theory of Functions* (Littlewood, 1944). Rado says in his review of Littlewood's text for the December 1945 issue of the *The Mathematical Gazette*,

> [The] sole aim [of the text] seems to be to make the reader share the author's delight in one of the most beautiful realms of mathematics.

One of the most referenced heuristics comes from Littlewood's text. He proposed three principles as guides for working in real analysis:

1. Every measurable set is nearly a finite union of intervals.

2. Every measurable function is nearly continuous.

3. Every convergent sequence of measurable functions is nearly uniformly convergent.

Royden (1988, p. 71), among many others, quotes Littlewood (1944, p.26) directly

> Most of the results are fairly intuitive applications of these ideas. If one of the principles would be the obvious means to settle the problem if it were 'quite' true, it is natural to ask if the 'nearly' is near enough, and for a problem that is actually solvable it generally is.

Littlewood's principles correspond to important theorems of Lebesgue (1902), Luzin[1] (1912), and Egorov (1911). However, the point of these maxims is not to highlight the theorems but rather to offer an approach to solving problems in real analysis. For example, suppose we are trying to establish a property that would be easy if the functions involved were continuous. Then we should attempt to prove the result using "nearly continuous," or measurable as measurable is "continuous except on a set of size less than ϵ." *By which theorem?* Many times, this "nearly" true is enough to prove the desired result. Littlewood's principles provide a very important approach to real analysis, so useful that, as in Royden (1988, Section 3.6) or Bichteler (1998, Section III.1), the principles have their own section.

Summary

We have looked at how arbitrary sets can be measured, generalizing the length of an interval. We then learned how to build an integral for successively larger classes of functions. The technique of creating a chain from simple functions to nonnegative to general functions is very useful and appears in many places in analysis. We then examined how much convergence properties were enhanced by Lebesgue's integral,

---

[1] Actually, in 1903, Lebesgue stated the result named for Luzin without proof; Vitali proved it in 1905.

essentially replacing the need for uniform convergence. The last section discussed how usually the properties we need are nearly true, so the theory isn't nearly as difficult as it first seems.

Today, there are two standard approaches to Lebesgue theory. The one we have used to develop measure is Carathéodory's method of 1914 that asks whether a set splits all other sets into pieces whose sizes add properly. For deeper and more general treatments, see, for example, Cohn (1980), Royden (1988), or Rudin (1976), and especially Bressoud's (2008) *A Radical Approach to Lebesgue's Theory of Integration,* which also shows Lebesgue's technique using inner and outer measures. Lebesgue's method is also illustrated in Boas's (1981) *A Primer of Real Functions.* The other commonly used approach first develops the Lebesgue integral and then defines the measure of a set to be the value of the integral of that set's characteristic function. This method is done very well in Chae's (1995) *Lebesgue Integration.* Burk's (2007) *A Garden of Integrals* is an excellent comparison of several different types of integrals.

There are numerous directions for further study. We could define an indefinite Lebesgue integral and investigate its properties. We could study spaces other than the real numbers to develop more general measure theory. We could generalize to infinite dimensional spaces of functions. We could return to Fourier series and convergence. There are exciting topics in every direction we look.

EXERCISES

3.1   Prove De Morgan's laws for sets.
  a) $(A \cup B)^c = A^c \cap B^c$
  b) $(A \cap B)^c = A^c \cup B^c$

3.2   Let M be a collection of sets and let

$$\mathcal{M} = \bigcap \{\mathcal{B} \text{ is an algebra and } M \subset \mathcal{B}\}$$

Prove: $\mathcal{M}$ is the smallest algebra containing M.

3.3   Show that the Borel σ-algebra $\mathcal{B}(\mathbb{R})$ also contains all the closed sets.

3.4   Prove the *monotonicity* of Lebesgue measure: if $A \subseteq B \in \mathfrak{M}$, then $\mu(A) \leq \mu(B)$.

3.5   If there is a set in $E \in \mathfrak{M}$ with finite measure, then $\mu(\emptyset) = 0$.

3.6   Let $X = [0, 1]$. Set $\mathcal{A}$ to be the collection of subsets $S \subseteq [0, 1]$ where either S or $S^c = [0, 1] - S$ is finite. Show
  a) $\mathcal{A}$ is an algebra.
  b) $\mathcal{A}$ is not a σ-algebra.

3.7   Show
  a) $\mu^*(\emptyset) = 0$
  b) $\mu^*(\{x\}) = 0$

3.8   Prove: μ^* is translation invariant. That is, $\mu^*(x + E) = \mu^*(E)$ for every $E \subset \mathbb{R}$ and $x \in \mathbb{R}$.

3.9   Prove: If E_1 and E_2 are measurable, then $E_1 \cap E_2$ is measurable.

3.10   Prove Theorem 3.13.

3.11   Suppose that $E \in \mathfrak{M}$. Show there exists an open set $O \supseteq E$ and a closed set $F \subseteq E$ such that $\mu(E - F) < \epsilon$ for any given $\epsilon > 0$.

3.12   Suppose $A \subset E$ and $\mu(E) = 0$. Then A is measurable and $\mu(A) = 0$.

3.13   Let A and B be two bounded, measurable subsets of $\mathbb{R}$. Prove

$$\mu(A \cup B) = \mu(A) + \mu(B) - \mu(A \cap B)$$

3.14   Give a class presentation on the *inclusion-exclusion principle.*

3.15 Prove Corollary 3.16, if a function satisfies the measurability conditions, then the set $f^{-1}(a)$ is measurable for every $a \in \mathbb{R}$.

3.16 Use Theorem 3.17 to prove that all polynomials are measurable.

3.17 Prove the statements of Theorem 3.18 for infima.

3.18 If f is measurable, then $|f| = \max\{f, -f\}$ is measurable. Give an example showing that the measurability of $|f|$ does *not* imply that of f.

3.19 Give an example to show that the measurability of the sets $\{f^{-1}(a)\}$ for every a does not imply that f is a measurable function.

3.20 If $a \leq b$, show that

$$\int \chi_{[a,b]} \, d\mu = b - a$$

3.21 If E is a bounded measurable set, then

$$\int_E d\mu = \int \chi_E \, d\mu = \mu(E)$$

3.22 Prove Theorem 3.26, the properties of Lebesgue integrals.

3.23 Suppose f is a bounded measurable function and E is a set of measure zero. Find $\int_E f \, d\mu$.

3.24 Prove Corollary 3.27,

$$\left| \int_E f \, d\mu \right| \leq \int_E |f| \, d\mu$$

3.25 Let

$$f(x) = \begin{cases} 1 & \dfrac{1}{2n} < x \leq \dfrac{1}{2n-1}, n \in \mathbb{N} \\ 0 & \text{otherwise} \end{cases}$$

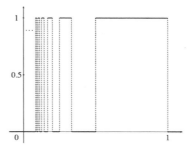

Show that both the Riemann and Lebesgue integrals exist and

$$\int_0^1 f(x) \, dx = \int_0^1 f \, d\mu = \ln(2)$$

3.26 Let $H(x)$ be the *unit step function*, $H(x) = 1$ if $x > 0$ and $H(x) = 0$ otherwise. Define *Zeno's staircase* by

$$Z(x) = \sum_{k=1}^{\infty} \frac{1}{2^k} \cdot H\left(x - \frac{k-1}{k}\right)$$

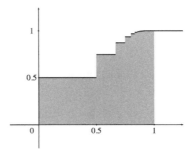

a) Determine if Z is Lebesgue integrable on $[0, 1]$. If so, find the value.

b) Determine if Z is Riemann integrable on $[0, 1]$. If so, find the value.

3.27 Let $S(x) = x^{-1/2}$ for $x \neq 0$ and $S(0) = 0$. Use n-truncations to show that $S \in \mathcal{L}([0, 1])$ to find the value of $\int_0^1 S \, d\mu$ as a limit.

3.28 Prove Theorem 3.30, the algebra of Lebesgue integrals for unbounded functions.

3.29   Suppose f is integrable on $[a, b]$ and $\epsilon > 0$. Prove there is a $\delta > 0$ such that for any subset $E \subset [a, b]$ with $\mu(E) < \delta$ we have

$$\left| \int_E f \, d\mu \right| < \epsilon$$

3.30   Determine whether

$$g_n(x) = \begin{cases} 1 & \dfrac{1}{2n} < x \le \dfrac{1}{2n-1} \\ 0 & \text{otherwise} \end{cases}$$

converges

a) pointwise on $[0, 1]$.

b) almost everywhere on $[0, 1]$.

c) in mean on $[0, 1]$.

3.31   Discuss the convergence properties of

$$D_n(x) = \begin{cases} n & x \in \mathbb{Q} \\ \dfrac{x^n}{n} & \text{otherwise} \end{cases}$$

for $x \in [0, 1]$.

3.32   Verify the claims in Example 3.11, part 1.

3.33   Verify the claims in Example 3.11, part 2.

3.34   Give an example of a sequence of functions that converges almost uniformly but not pointwise on $\mathbb{R}$.

3.35   Use the characterization of measurable functions as limits almost everywhere of step functions and Egorov's theorem to prove Theorem 3.21, Luzin's theorem.

3.36   Suppose that f is Lebesgue integrable and $|f(x)| < 1$ for all $x \in [-1, 1]$. Show that

$$\int_{[-1,1]} f^n \, d\mu = 0$$

3.37   Determine the value of

$$\int_0^1 \left(\sum_{n=1}^{\infty} \frac{x^n}{n} \right) d\mu$$

3.38   Calculate the value of

$$\int_0^1 \left(\sum_{n=0}^{\infty} \frac{x^n}{n!} \right) d\mu$$

with two different methods. First, using Lebesgue's dominated convergence theorem and, second, by recognizing the summation as a standard function.

3.39   Verify the claims of Example 3.13.

3.40   Define h_n by $h_n(x) = nxe^{-nx^2}$.

a) Compute

$$\int_0^1 \lim_{n \to \infty} h_n \, d\mu$$

Hint: look at graphs of h_n.

b) Compute

$$\lim_{n \to \infty} \int_0^1 h_n \, d\mu$$

c) Explain the results in relation to Lebesgue's dominated convergence theorem.

3.41   Let $f_n(x) = (n+1)x^n$ on $[0, 1]$.

a) Compute

$$\int_{[0,1]} \liminf_{n \to \infty} f_n \, d\mu$$

b) Compute

$$\liminf_{n \to \infty} \int_{[0,1]} f_n \, d\mu$$

c) Explain the results in relation to Fatou's lemma.

3.42   Give a class presentation on Lebesgue's life and mathematics.

3.43   State the three theorems corresponding to Littlewood's three principles.

3.44   Give a class presentation on Littlewood's life and mathematics.

3.45   Who said,

In my opinion, a mathematician, in so far as he is a mathematician, need not preoccupy himself with philosophy—an opinion, moreover, which has been expressed by many philosophers.

INTERLUDE: THE SET OF RATIONAL NUMBERS IS VERY LARGE AND VERY SMALL

The structure of the real line is very rich. We can learn a good deal by looking at special subsets such as the rational numbers $\mathbb{Q}$. The rational numbers make a very interesting set in their own right. The rationals are *dense* in the reals; that is, the closure of $\mathbb{Q}$ is $\mathbb{R}$. Alternate forms of density are: every interval in $\mathbb{R}$ contains a rational number, or between any two real numbers lies a rational number. The rationals are countable, while the reals are uncountable. So there are a great deal fewer rational numbers than reals but yet many more rationals than integers. The rationals form a very small part of the real numbers; we can "cover" them all with a collection of tiny sets. Ultimately, the question "How large is $\mathbb{Q}$?" has very different answers depending on the context.

$\mathbb{Q}$ is Countably Infinite

Define *cardinality* as the number of elements in a set and denote the cardinality of S by $\#(S)$. Clearly cardinality is nonnegative—sets are not allowed to "owe the universe elements." *What is the cardinality of the empty set?* The cardinality of the natural numbers is infinite and is written as $\#(\mathbb{N}) = \aleph_0$ (read as "aleph null"). Two sets have the same cardinality if and only if there is a one-to-one, onto mapping between them. Since $\mathbb{N} \subset \mathbb{Q}$, then $\#(\mathbb{N}) \leq \#(\mathbb{Q})$.

Theorem I.1 $\mathbb{Q}$ *is countably infinite and* $\#(\mathbb{Q}) = \aleph_0$.

Proof: The Schröder-Bernstein theorem states that if $\#(A) \leq \#(B)$ and $\#(B) \leq \#(A)$, then $\#(A) = \#(B)$, so we need only show that $\#(\mathbb{Q}) \leq \#(\mathbb{N})$.

Write an array containing all the rational numbers by putting all fractions with denominator 1 in the first row of an infinite matrix, all fractions with denominator 2 in the second row of an infinite matrix, and so forth.

$$
\begin{array}{ccccc}
\frac{1}{1}(1) & \frac{2}{1}(2) & \frac{3}{1}(4) & \frac{4}{1}(7) & \cdots \\[4pt]
\frac{1}{2}(3) & \frac{2}{2}(5) & \frac{3}{2}(8) & \frac{4}{2}(12) & \cdots \\[4pt]
\frac{1}{3}(6) & \frac{2}{3}(9) & \frac{3}{3}(13) & \frac{4}{3}(18) & \cdots \\[4pt]
\frac{1}{4}(10) & \frac{2}{4}(14) & \frac{3}{4}(19) & \frac{4}{4}(25) & \cdots \\[4pt]
\vdots & \vdots & \vdots & \vdots & \ddots
\end{array}
$$

Going down reverse diagonals, map each rational to the integer listed in the array. For example, $r(1/1) = 1$, $r(2/1) = 2$, $r(1/2) = 3$, $r(3/1) = 4$, $r(2/2) = 5$, etc. Since each rational has an infinite number of representations, the mapping r is not one-to-one to $\mathbb{N}$. However, removing the duplicates gives a restriction of r that is one-to-one and onto a subset of $\mathbb{N}$. Hence, $\#(\mathbb{Q}) \leq \#(\mathbb{N})$. Therefore, by the Schröder-Bernstein theorem, $\#(\mathbb{Q}) = \#(\mathbb{N}) = \aleph_0$. ∎

We see that the number of rationals is equal to the number of natural numbers.

See Calkin & Wilf (2000) for a very clever binary-tree-based argument showing the rationals are countable.

$\mathbb{Q}$ is Dense in $\mathbb{R}$

Density describes how "thickly" one set is inside another. A set A is *dense* in a set B if the closure $\bar{A} = B$. Recall that the closure of a set is the set itself together with all its accumulation points. A point a is an accumulation point of A if every neighborhood of a contains another point of A. Since neighborhoods on the real line contain open intervals, we can rephrase "$\mathbb{Q}$ is dense in $\mathbb{R}$" succinctly as "between any two reals, there lies a rational." Hankel called dense sets "sets in close order."

Theorem I.2 $\mathbb{Q}$ *is dense in* $\mathbb{R}$.

Proof: By our comments above, we need only show that between any two real numbers there must be a rational. Let $x < y \in \mathbb{R}$ be given. Since $y - x > 0$, there is a positive integer n such that $y - x > 1/n$. *Why?*

Choose $k = \min\{j \in \mathbb{N} \mid j > nx\}$. *How do we know such k exists?*   Then $x < k/n$.

Suppose $y \leq k/n$. Then

$$\frac{k}{n} - \frac{1}{n} = \frac{k-1}{n} \geq y - \frac{1}{n} > y - (y - x) = x$$

which is a contradiction. *Why?* Therefore, $k/n < y$.

Thus, we have

$$x < \frac{k}{n} < y$$

∎

Would $n = 1 + \lfloor 1/(y - x) \rfloor$ and $k = \lceil nx \rceil$ work in the proof above?

Since $\mathbb{Q}$ is dense in $\mathbb{R}$, the rationals are spread throughout the reals. Let $x \in \mathbb{R}$. Then there is a rational r_n in each interval $(x, x + 1/n)$. So for any real number x we can find a sequence of rationals converging to x.

The Set of Real Numbers is Uncountable

How large is the set of real numbers? Uncountably infinite.

Theorem I.3 $\mathbb{R}$ *is uncountably infinite.*

Proof: We'll show that assuming $\mathbb{R}$ is countable leads to a contradiction. The construction following is called a *Cantor diagonalization argument*. We'll focus on the reals in $(0, 1)$ for simplicity.

Suppose that $(0, 1)$ is countable. Write $(0, 1)$ as

$$(0, 1) = \{x_1, x_2, x_3, \ldots\}$$

Any $x \in (0, 1)$ has a binary expansion. List these expansions in an array.

$$
\begin{aligned}
x_1 &= 0 . \mathbf{1}\ 0\ 0\ 0\ 1\ \ldots \\
x_2 &= 0 . 0\ \mathbf{0}\ 1\ 0\ 1\ \ldots \\
x_3 &= 0 . 1\ 1\ \mathbf{0}\ 1\ 0\ \ldots \\
x_4 &= 0 . 0\ 1\ 1\ \mathbf{1}\ 0\ \ldots \\
x_5 &= 0 . 0\ 1\ 1\ 0\ \mathbf{1}\ \ldots \\
&\ \ \vdots
\end{aligned}
$$

$$
x^* = 0 . \mathbf{0}\ \mathbf{1}\ \mathbf{1}\ \mathbf{0}\ \mathbf{0}\ \ldots
$$

Construct the number x^* by setting the nth bit of x^* equal to $1 - n$th bit of x_n. Since the nth bit of x^* differs from the nth bit of x_n for every n, we have that x^* does not appear in the list of all numbers—a contradiction. Therefore $(0, 1)$ is uncountable. ∎

We write $\#(\mathbb{R}) = \mathfrak{c}$.

The *continuum hypothesis* states *there is no set whose size is strictly between* $\mathbb{N}$ *and* $\mathbb{R}$. If we set $\aleph_1$ to be the first infinite cardinal greater than $\aleph_0$, then the continuum hypothesis becomes $\aleph_1 = \mathfrak{c}$. [See Aigner & Ziegler (2001, Chapter 15).] *What is the cardinality of the set of irrational numbers?* Cantor first posed this question in the 1890s. Hilbert put the hypothesis first on the famous list of unsolved problems from his address to the International Congress of Mathematics in 1900. The work of Gödel (from 1940) and Chen (from 1963) shows the continuum hypothesis is independent of the standard axioms of set theory; that is, it can neither be proved nor disproved. Currently, mathematicians are divided; many believe it true, others think it false. The question is still open today.

$\mathbb{Q}$ is a Null Set

The rationals are countably infinite and dense in the reals. But there are no more rationals than integers in that their cardinalities are the same. The integers are sparsely distributed in the reals. Are the rationals also sparsely distributed in some sense? Yes. Since the rationals are countable and the reals are uncountable, the rationals "can't take up much space" on the real line.

A set $A \subseteq \mathbb{R}$ is a *null set* or a *set of measure zero* if it can be covered by a collection of open intervals having total length as small as desired. The rationals are a null set.

Theorem I.4 $\mathbb{Q}$ *is a null set.*

Proof: Let $\epsilon > 0$.

Since $\mathbb{Q}$ is countable, list the elements as

$$
\mathbb{Q} = \{r_1, r_2, r_3, \ldots\}
$$

Construct a collection of open intervals that forms an *open cover* of $\mathbb{Q}$. Set $O_1 = (r_1 - \epsilon/4, r_1 + \epsilon/4)$. Then the length of O_1 is $\epsilon/2$. Set $O_2 = (r_2 - \epsilon/8, r_2 + \epsilon/8)$.

Then the length of O_2 is $\epsilon/4$. Continue in this fashion to generate a sequence of open intervals such that

1. $r_n \in O_n = (r_n - \epsilon/2^{n+1}, r_n + \epsilon/2^{n+1})$.

2. the length of O_n is $\epsilon/2^n$.

Since each r_n is in the interval O_n, it follows that $\mathbb{Q} \subset \bigcup_n O_n$.
The total length of the open cover $\mathcal{C} = \{O_n\}$ is

$$\ell(\mathcal{C}) = \sum_{n=1}^{\infty} \ell(O_n) = \sum_{n=1}^{\infty} \frac{\epsilon}{2^n} = \epsilon \cdot \sum_{n=1}^{\infty} \frac{1}{2^n} = \epsilon$$

∎

In the language of Lebesgue measure, $\mathbb{Q}$ is a set of measure zero. Recall that a property holds *almost everywhere* if the set where it fails to hold has measure zero. Dirichlet's monster function, $D(x)$ equals 1 if $x \in \mathbb{Q}$ and 0 otherwise, is equal to zero almost everywhere. Hence the Lebesgue integral of D is zero.

The count of individual rational numbers is huge—there are infinitely many. Rationals are dense in the reals—they are everywhere. However, if we consider the portion of the real line formed by the rationals, it is significantly insignificant—the rationals have measure zero. So $\mathbb{Q}$ is both very large and very small.

CHAPTER 4

SPECIAL TOPICS

It is not the knowledge, but the act of learning, not the possession but the act of getting there, which grants the greatest enjoyment. When I have clarified and exhausted a subject, then I turn away from it, in order to go into the darkness again.

—Gauss

Introduction

Now that we've studied analysis, let's turn to a set of special topics that have connections to elementary calculus. Each subject expands on one basic concept of analysis, making an advanced use of a simple idea. As we go into each topic, look for the link back to the simple idea from calculus. However, also look for deeper connections than "the logistic model appears in the AP Calculus syllabus;" ask *Why? What aspect of elementary calculus is featured and developed?* Especially consider possible uses of each of the special topics in the elementary calculus classroom.

Opening quotation is from an 1808 letter of Gauss to Bolyai. See, e.g., Dunnington's *Carl Friedrich Gauss: Titan of Science* (2004, p. 416), Math. Assoc. of America, Washington, DC.

4.1 MODELING WITH LOGISTIC FUNCTIONS—NUMERICAL DERIVATIVES

Pierre Verhulst, a Belgian mathematician, wrote "Notice sur la loi que la population poursuit dans son accroissement" ("Note on the law of population growth") (Verhulst, 1838) arguing that population growth could not be unlimited as Malthus had proposed in 1798 in *An Essay on the Principle of Population*. [See Malthus (1826).] Verhulst wrote

> If the population increases by geometric progression, we would have the equation $dp/dt = mp$. However, as the rate of population growth is slowed by the very increase in the number of inhabitants, we must subtract from mp an unknown of p [Verhulst, 1838, p. 186 (translation)].

Verhulst investigated growth models based on the differential equation

$$\frac{dp}{dt} = mp - \phi(p)$$

where $p(t)$ represents the population count at time t. He tried $\phi(p) = np^2$, np^3, np^4, and $n\ln(p)$, successively, settling on np^2 claiming the "differences between the populations calculated and those furnished by observation were approximately the same" and he "chose the most simple." [See the translation (Verhulst, 1838, p. 186).] Figure 4.1 shows typical logistic curves superimposed on a slope field of the logistic differential equation. Even though developed over 150 years ago, the logistic model is still widely used in very diverse fields. For example, four current papers are:

- "Farmers' satisfaction with aquaculture—A logistic model in Vietnam" by Duc (2008).

- "Evaluation of logistic and polynomial models for fitting sandwich-ELISA calibration curves" by Herman et al. (2008).

- "The logistic (Verhulst) model for sigmoid microbial growth curves revisited" by Peleg et al. (2007).

- "Variation in and correlation between intrinsic rate of increase and carrying capacity" by Underwood (2007).

The name "logistic" is an interesting choice; the word's origin is Greek: $\lambda o \gamma \iota \sigma \tau \iota \kappa o \varsigma$ (logistikos), meaning "skilled in calculation." The term appears to have been first used by the geographer Edward Wright in 1599 but for logarithmic curves rather than sigmoid or "s-shaped" functions. In 1844, Verhulst wrote, "Nous donnerons le terme de logistique à cette courbe," naming his model a "logistic curve" (Verhulst, 1844). It appears that Verhulst coined the term based on the log-like properties of his function, but he never explained his choice. [See Shulman (1998) for an interesting discussion of the origins of the term logistic and on using history pedagogically in the mathematics classroom.]

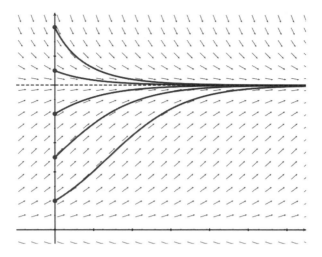

Figure 4.1 Logistic Functions and Slope Field

The Logistic Equation

The logistic model has come to be a standard part of elementary calculus since it is widely applied and it's relatively easy to solve the differential equation. The curve has a nice collection of properties for the beginning calculus student to find with easy calculations: boundedness, no extreme values, one inflection point, different limits at plus and minus infinity. *Verify these!*

We solve this separable differential equation in the standard way. Divide both sides of the logistic differential equation by $mp - np^2$ assuming $p \neq 0$ or m/n; then use a partial fractions expansion to have

$$\frac{1}{mp - np^2} \frac{dp}{dt} = \left(\frac{1}{mp} + \frac{m}{n(m - np)} \right) \frac{dp}{dt} = 1$$

Integrate both sides with respect to t. Then

$$\frac{\ln(p)}{m} - \frac{\ln(m - np)}{m} = t + C$$

Solve this expression for p assuming that $0 < p \leq m/n$. Hence

$$p = \frac{m/n}{1 + (e^{-Cm}/n) \cdot e^{-mt}}.$$

Do the calculation for $p > m/n$. We can rewrite the expression above using an initial value $p(t_i) = p_i$ to arrive at the *logistic equation*

$$p(t) = \frac{m/n}{1 + e^{mt_i} \left[(m/n)/p_i - 1 \right] e^{-mt}}$$

The notation becomes much more compact if we let $m/n = M$ be the *carrying capacity*, $m = r$ be the relative growth rate, and $b = e^{rt_i}((M/p_i) - 1)$ be a constant depending on M, r, and the initial values. Then the logistic equation becomes

$$p(t) = \frac{M}{1 + be^{-rt}}$$

Write the logistic differential equation using r and M instead of m and n.

Fitting Data with a Logistic Model

The logistic model was rediscovered by Pearl and Reed in 1920, then again by Lotka in 1925, and yet again by Volterra in 1928. Pearl and Reed authored a series of papers over 20 years on logistic models; their last paper, "The Logistic Curve and the Census Count of 1940" (Pearl et al., 1940), fit a logistic model to the population of the United States. Pearl, Reed, and Kish used the census data in Table 4.1 to derive the formula

$$y = \frac{184.00}{1 + 66.69\,e^{-0.0322x}}$$

(y is in millions, x is years from 1780) with a technique called *successive least squares approximation* (Pearl et al., 1940, p. 488); their logistic fit is shown in Figure 4.2.

Table 4.1   US Population, 1790–1940. From (Pearl et al., 1940)

Year	US Population (millions)
1790	3.929
1800	5.308
1810	7.240
1820	9.638
1830	12.866
1840	17.069
1850	23.192
1860	31.443
1870	38.558
1880	50.156
1890	62.948
1900	75.995
1910	91.972
1920	105.711
1930	122.775
1940	131.410

Now we can get a similar result,

$$y = \frac{180.27}{1 + 71.53\,e^{-0.0330x}} + 0.672$$

in under a minute using a TI-Nspire calculator's logistic regression function. However, there are only 16 data points. A large data set, say the population of the United States from 1790 to 2010 given yearly, would overwhelm our calculator, not to mention our fingers. Let's investigate a method, different from Pearl's, for approximating a logistic fit that uses two important techniques: transforming data and linear least squares fitting. We'll use Pearl, Reed, and Kish's data from Table 4.1.

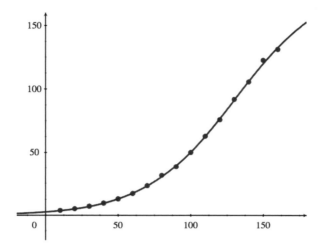

Figure 4.2   Pearl, Reed, and Kish's Fit to the U.S. Population

First, transform Verhulst's logistic differential equation. If we divide both sides by p, the right side becomes linear.

$$\frac{1}{p}\frac{dp}{dt} = m - np$$

Linear functions are easy to fit to data. How do we handle p'/p? We need to estimate the derivative p' numerically. Our main tool will be Taylor polynomials.

The derivative of $f'(a)$ is the limit of the difference quotients

$$\frac{\Delta y}{\Delta x} = \frac{f(a+h) - f(a)}{h}$$

Expand the expression $f(a+h) - f(a)$ in a Taylor polynomial in powers of h to order 2 centered at $h = 0$ obtaining

$$f(a+h) - f(a) \approx f'(a)\,h + \frac{1}{2}f''(a)\,h^2$$

What are we assuming about f and h? Divide by h.

$$\frac{f(a+h) - f(a)}{h} \approx f'(a) + \frac{1}{2}f''(a)\,h$$

If f'' is bounded and h is small enough, we can say

$$\frac{f(a+h) - f(a)}{h} \approx f'(a)$$

This formula is called a *forward difference.* To use the forward difference approximation of the derivative with the population data, set $a = t_n$ and $a + h = t_{n+1}$. Then $h = t_{n+1} - t_n$, $f(a) = p_n$, and $f(a + h) = n_{n+1}$, so that

$$p'(t_n) \approx \frac{p_{n+1} - p_n}{t_{n+1} - t_n}$$

A *backward difference* is developed analogously. Expand $f(a) - f(a - h)$ in a Taylor polynomial in powers of h to order 2 centered at $h = 0$, then divide by h, obtaining

$$\frac{f(a) - f(a-h)}{h} \approx f'(a) - \frac{1}{2} f''(a)\, h$$

To use the backward difference approximation of the derivative with our population data, set $a = t_1$ and $a - h = t_0$. Then $h = x_1 - x_0$, $f(a) = p_1$, and $f(a - h) = p_0$. Thus

$$p'(t_n) \approx \frac{p_n - p_{n-1}}{t_n - t_{n-1}}$$

Both the forward and backward differences have errors proportional to the first power of h. We say these methods are $O(h)$.

Taking the average of the forward and backward differences leads to the *symmetric difference*, sometimes called the *three-point formula.* Once again find the Taylor polynomial, but this time for $f(a + h) - f(a - h)$. We have

$$\frac{f(a+h) - f(a-h)}{2h} \approx f'(a) + \frac{1}{6} f'''(a)\, h^2$$

Hence, for constant h or equally spaced t_n's,

$$p'(t_n) \approx \frac{p_{n+1} - p_{n-1}}{t_{n+1} - t_{n-1}}$$

Write out the details! There is no $f''(a)\, h$ term—the symmetric difference error is $O(h^2)$. There are more intricate formulas for estimating derivatives from data (*Look up the five-point formula!*) that we could use, but the symmetric difference will work well for us.

Next, we transform our data from (t, p) to $(p, p'/p)$ pairs using a symmetric difference approximation for p'. Since $h = 10$ for all points, p'/p entries are found with

$$\frac{p'}{p} = \frac{1}{p_n} \frac{p_{n+1} - p_{n-1}}{20}.$$

Why are there no entries in the first and last rows of the p'/p column? The line of

Table 4.2   First Data Transformation

t	p	p	p'/p
1790	3.929	3.929	
1800	5.308	5.308	0.03119
1810	7.240	7.240	0.02990
1820	9.638	9.638	0.02920
1830	12.866	12.866	0.02888
1840	17.069	17.069	0.03023
1850	23.192	23.192	0.03098
1860	31.443	31.443	0.02444
1870	38.558	38.558	0.02428
1880	50.156	50.156	0.02431
1890	62.948	62.948	0.02052
1900	75.995	75.995	0.01909
1910	91.972	91.972	0.01614
1920	105.711	105.711	0.01458
1930	122.775	122.775	0.01046
1940	131.410	131.410	

best fit to the $(p, p'/p)$ transformed data of Table 4.2 is

$$y = 0.0318 - 0.0001696x$$

The least squares line is easily found with a calculator, a spreadsheet, or a computer algebra system such as Maple. The coefficients of the linear fit tell us $m = 0.0318$ and $n = 0.0001696$. Then $r = 0.0318$ and $M = 187.58$. So

$$p(t) = \frac{187.58}{1 + be^{-0.0318t}}$$

How do we choose an initial value to use to determine b? There's no reason to prefer one point over another, although more recent values are probably more accurate due to improved census methods. We apply another transformation to the data. First, invert the logistic equation, then subtract $1/M$ from both sides.

$$p(t) = \frac{M}{1 + be^{-rt}}$$

$$\frac{1}{p} = \frac{1 + be^{-rt}}{M} = \frac{1}{M} + \frac{b}{M}e^{-rt}$$

$$\frac{1}{p} - \frac{1}{M} = b\frac{e^{-rt}}{M}$$

Now define $y = 1/p - 1/M$ and $x = e^{-rt}/M$ to have

$$y = bx$$

Table 4.3 Second Transformation of U.S. Population Data

year	p	x	y
1790	3.929	0.00388	0.24919
1800	5.308	0.00282	0.18306
1810	7.240	0.00205	0.13279
1820	9.638	0.00149	0.09842
1830	12.866	0.00109	0.07239
1840	17.069	0.00079	0.05325
1850	23.192	0.00058	0.03779
1860	31.443	0.00042	0.02647
1870	38.558	0.00030	0.02060
1880	50.156	0.00022	0.01461
1890	62.948	0.00016	0.01055
1900	75.995	0.00012	0.00783
1910	91.972	0.00009	0.00554
1920	105.711	0.00006	0.00413
1930	122.775	0.00005	0.00281
1940	131.410	0.00003	0.00228

Examine the transformed data in Table 4.3. (Remember $t = year - 1780$.) A second least squares linear fit gives us $b = 64.74$. Our logistic fit to the U.S. population from 1790 to 1940 is

$$p(t) = \frac{187.58}{1 + 64.74\,e^{-0.0318t}}$$

Our values are close to Pearl, Reed, and Kish's; we plot their fit, our fit, and the data in Figure 4.3. We see the different logistic models are extremely close to each other and fit the data well. Our curve is just above Pearl, Reed, and Kish's logistic fit of 1940.

Our method relied on numerical approximations to the derivative. We used Taylor polynomials not only to find a formula for the approximation but also to examine the error involved. There has been a theme in the computations: transformation to linear. Linear functions are easy to work with and easy to find, so we manipulated our data to fit linear models. Many calculations in science use linearization in some form.

For an analysis of our technique, see Bauldry (1997), where Maple was used to fit a logistic model to the U.S. population from 1790 to 1990. The data set for the U.S. population for each year from 1790 to 2008 can be downloaded from

www.mathsci.appstate.edu/~wmcb/IA/Data/USPop1790to2008.txt

4.2 NUMERICAL QUADRATURE

Calculating areas is one of the seminal problems of calculus that dates to the ancient Greeks. Eudoxes and Archimedes, among others, began a long line of investigations into an area that would become integral calculus. The term *quadrature* is very old;

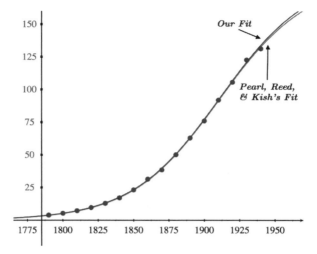

Figure 4.3 Logistic Fit Comparisons

Bailey's 1731 *Universal Etymological English Dictionary*[1] lists

QUADRATURE [*quadratura*, L] the making of a thing fquare, or the finding a fquare equal to the area of any figure given.

QUADRATURE *of Curves* [in the higher *Geometry*] is the meafuring of their area, or the finding of a rectilinear fpace, equal to a curvilinear fpace.

Current dictionaries, such as the *Oxford English Dictionary* or *Webster's Revised Unabridged Dictionary*, give the same primary meaning to "quadrature." The sense of the word we intend to explore matches Maple's "Mathematics and Engineering Dictionary," which states:

numerical quadrature. the evaluation of a definite integral by a formula involving weighted sums of function values at given points.

First, we will consider the standard formulas from elementary calculus, left endpoint, trapezoid, midpoint, and Simpson's rules, building them by approximating the integrand function with polynomials. Then we will look at Gaussian quadrature and its properties. To simplify our calculations, we'll assume we have continuous functions with as many derivatives as needed.

Let's set the stage. The example integral we will use, shown in Figure 4.4, is

$$\int_0^3 \cos(8\sin(x-1)) + 1 \, dx$$

A computer algebra system gives the value of this integral to 10 places as 3.4351935056. This function, which all of Maple, Mathematica, and TI-Nspire agree has no elemen-

[1] See www.google.com/books?id=o-gIAAAAQAAJ&printsec=frontcover&dq=quadrature#PPT642,M1

tary antiderivative, is one of the standards commonly used to test numeric integration methods listed in Zwillinger (1992, p. 273).

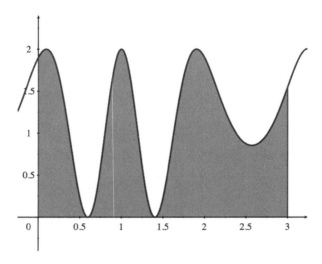

Figure 4.4   Our Test Integral

To approximate $\int_a^b f(x)\,dx$, start by partitioning the interval $[a, b]$ into n equal subintervals. Then $h = (b-a)/n$. Our partition is $P = \{x_k = a + kh \mid k = 0, \dots, n\}$. For each k, set $y_k = f(x_k) = f(a + kh)$.

The Left Endpoint Rule

The 0th-degree approximation to a function is a horizontal line matching a particular function value. Choose the left endpoint of each interval to give $y_k = f(x_{k-1}) = f(a + (k-1)h)$. Then

$$\int_a^b f(x)\,dx \approx L_n = h\left(y_0 + y_1 + \cdots + y_{n-1} + y_n\right)$$

Figure 4.5 shows a left endpoint approximation with six panels; $L_6 = 3.5694$.

An approximation without an error bound is of no use, so we turn our focus to analyzing the error. Use the first-order Taylor polynomial centered at x_{k-1} to write

$$f(x) = f(x_{k-1}) + f'(c_k)(x - x_{k-1})$$

where $c_k \in (x_{k-1}, x_k)$. Now integrate both sides from x_{k-1} to x_k to find

$$\int_{x_{k-1}}^{x_k} f(x)\,dx = f(x_{k-1})h + f'(c_k) \cdot \frac{h^2}{2}$$

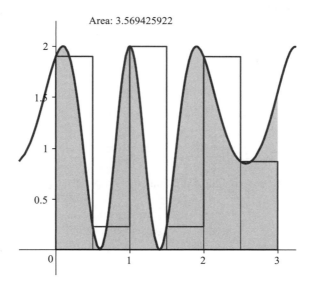

Figure 4.5   Left Endpoint Approximation with Six Panels

Summing for $k = 1, \ldots, n$ gives

$$\int_{x_0}^{x_n} f(x)\,dx = L_n + \frac{h^2}{2} \cdot \sum_{k=1}^{n} f'(c_k)$$

Let $M_1 = \max_{[a,b]} |f'(x)|$. Then the absolute value of the maximum error is

$$\varepsilon(L_n) = \left| \int_a^b f(x)\,dx - L_n \right| \le \frac{h^2}{2} \cdot n \cdot M_1 = \frac{(b-a)^2}{2} \cdot M_1 \cdot \frac{1}{n}$$

We have proven

Theorem 4.1 *Let f be differentiable on (a,b) and continuous on $[a,b]$. Also, let $M_1 \ge |f'(x)|$ for all $x \in [a,b]$. Then*

$$\varepsilon(L_n) \le \frac{(b-a)^2}{2} \cdot M_1 \cdot \frac{1}{n}$$

Check that $\varepsilon(L_6) = 0.13423$ satisfies this bound.

The Trapezoid and Midpoint Rules

The first degree approximation to f is linear. The line through (x_{k-1}, y_{k-1}) and (x_k, y_k) forms a trapezoid, so the area enclosed is $A = h(y_k + y_{k-1})/2$. The total area is the sum of the trapezoids.

$$
\begin{aligned}
T_n &= \sum_{k=1}^{n} \frac{1}{2} \left(y_{k-1} + y_k \right) h \\
&= \frac{h}{2} \left(y_0 + 2y_1 + 2y_2 + \cdots + 2y_{n-1} + y_n \right)
\end{aligned}
$$

Figure 4.6 shows a trapezoid approximation with six panels; $T_6 = 3.48108$.

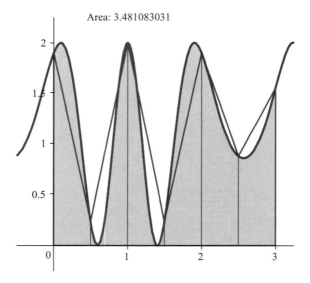

Figure 4.6   Trapezoid Rule with Six Panels

To generate an error bound for the trapezoid rule, we'll use a clever technique that goes back to Peano. Let m_k be the midpoint of the interval (x_{k-1}, x_k). Integrate the expression $P_k = \int_{x_{k-1}}^{x_k} (t - m_k) f'(t) \, dt$ by parts (*Exercise!*) to see

$$
\begin{aligned}
P_k &= \frac{h}{2} \left(f(x_{k-1}) + f(x_k) \right) - \int_{x_{k-1}}^{x_k} f(x) \, dx \\
&= \frac{h}{2} \left(y_{k-1} + y_k \right) - \int_{x_{k-1}}^{x_k} f(x) \, dx
\end{aligned}
$$

Now integrate the expression $Q_k = (1/2) \int_{x_{k-1}}^{x_k} \left[(h/2)^2 - (t - m_i)^2 \right] f''(t) \, dt$ by parts (*Exercise!*) to see that $Q_k = P_k$. Sum Q_k for $k = 1, \ldots, n$ to have

$$\sum_{k=1}^{n} Q_k = \sum_{k=1}^{n} P_k = T_n - \int_a^b f(x) \, dx$$

Let $M_2 = \max_{[a,b]} |f''(x)|$. Then

$$|Q_k| \le \frac{M_2}{2} \int_{x_{k-1}}^{x_k} \left| (h/2)^2 - (t - m_i)^2 \right| \, dt = \frac{M_2}{2} \cdot \frac{h^3}{6}$$

Verify the integration! Since $\varepsilon(T_n) = |\sum Q_k| \le \sum |Q_k|$, we have

$$\varepsilon(T_n) \le \frac{M_2}{2} \cdot \frac{h^3}{6} \cdot n = \frac{(b-a)^3}{12} \cdot M_2 \cdot \frac{1}{n^2}$$

We have proven

Theorem 4.2 *Let f be a function with a continuous second derivative on (a, b). Also, let $M_2 \ge |f''(x)|$ for all $x \in [a, b]$. Then*

$$\varepsilon(T_n) \le \frac{(b-a)^3}{12} \cdot M_2 \cdot \frac{1}{n^2}$$

Check that $\varepsilon(T_6) = 0.04589$ satisfies this bound.

Comparing the two rules, we see that doubling n halves the error in the left endpoint rule but quarters the error in the trapezoid rule. Since $\varepsilon(T_n)$ is bounded by the second derivative of f, the trapezoid rule must give the exact value for linear functions.

What about averaging horizontally rather than vertically? Let's use the midpoint of each interval as the point to compute the height of a rectangle, so the kth approximation to f is

$$y_k = f\left(\frac{x_{k-1} + x_k}{2} \right)$$

The total area is just the sum of the n rectangles. Thus

$$\int_a^b f(x) \, dx \approx \mathrm{MP}_n = h \sum_{k=1}^{n} f\left(\frac{x_{k-1} + x_k}{2} \right).$$

Figure 4.7 shows a midpoint rule approximation with six panels; $\mathrm{MP}_6 = 3.39825$.

We return to using Taylor polynomials to develop error bounds. Again, let m_k be the midpoint of the interval (x_{k-1}, x_k). Expanding f in a Taylor polynomial with remainder gives

$$f(x) = f(m_k) + f'(m_k)(x - m_k) + f''(c_k)(x - m_k)^2$$

for some point $c_k \in (x_{k-1}, x_k)$. Integrate from x_{k-1} to x_k. We have

$$\int_{x_{k-1}}^{x_k} f(x) \, dx = f(m_k)h + \frac{1}{2} \int_{x_{k-1}}^{x_k} f''(c_k)(x - m_k)^2 \, dx$$

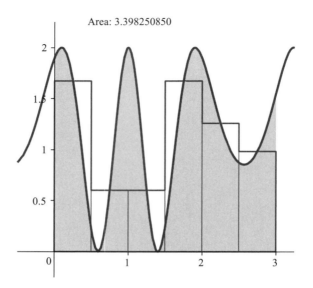

Figure 4.7 Midpoint Rule with Six Panels

because $\int_{x_{k-1}}^{x_k} (x - m_k)\, dx = 0$. Set $M_2 = \max_{[a,b]} |f''(x)|$. Then each integral with $f''(c_k)$ can be bounded by

$$\left| \frac{1}{2} \int_{x_{k-1}}^{x_k} f''(c_k)(x - m_k)^2\, dx \right| \leq \frac{M_2}{2} \int_{x_{k-1}}^{x_k} (x - m_k)^2\, dx = \frac{M_2}{2} \cdot \frac{h^3}{12}$$

Summing and using the integrals' bounds yields

$$\varepsilon(\mathrm{MP}_n) = \left| \int_a^b f(x)\, dx - \mathrm{MP}_n \right| \leq \frac{M_2}{2} \cdot \frac{h^3}{12} \cdot n = \frac{(b-a)^3}{24} \cdot M_2 \cdot \frac{1}{n^2}$$

We have proven

Theorem 4.3 *Let f be a function with a continuous second derivative on (a, b). Also, let $M_2 \geq |f''(x)|$ for all $x \in [a, b]$. Then*

$$\varepsilon(\mathrm{MP}_n) \leq \frac{(b-a)^3}{24} \cdot M_2 \cdot \frac{1}{n^2}$$

Check that $\varepsilon(\mathrm{MP}_6) = 0.039643$ satisfies this bound. Note that in general $\varepsilon(\mathrm{MP}_n)$ is half of $\varepsilon(T_n)$. It's interesting to observe that shifting from the left endpoint to the midpoint as the place to measure function height, we go from error bounded by $1/n$ to $1/n^2$. That's quite an improvement!

Simpson's Rule

The last quadrature method based on polynomial approximation that we will study is Simpson's rule.[2] Since a parabola requires three points to be uniquely determined, n must be even. Fit a parabola p to the three points (x_k, y_k), (x_{k+1}, y_{k+1}), and (x_{k+2}, y_{k+2}). Then integrate p from x_k to $x_k + 2h$. The resulting area under p is

$$\left(\frac{1}{3}y_k + \frac{4}{3}y_{k+1} + \frac{1}{3}y_{k+2} \right) h$$

The total area S_n for Simpson's rule is the sum of the $n/2$ parabolic regions

$$S_n = \frac{h}{3} \left(y_0 + 4y_1 + 2y_2 + 4y_3 + 2y_4 + \cdots + 4y_{n-1} + y_n \right)$$

Figure 4.8 shows a Simpson's rule approximation with six panels; $S_6 = 2.76630$. Note, using six panels approximates the function with three parabolas.

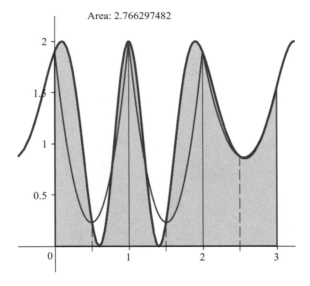

Figure 4.8   Simpson's Rule Approximation with Six Panels

The error estimate for Simpson's rule shows quite an improvement over both the midpoint and trapezoid rules. Initially, the reduction in error comes as a surprise since

[2]Curiously, Simpson's rule was developed by Newton, and Newton's method (following tangent lines to find roots) was invented by Simpson—mathematical nomenclature can be quite odd. See O'Connor & Robertson (2008).

Simpson's rule can be written as the weighted average

$$S_n = \frac{2}{3} \operatorname{MP}_n + \frac{1}{3} T_n$$

We follow the development from Apostol (1969, p. 605).

Theorem 4.4 *Let f be a function with a continuous fourth derivative on (a, b). Also, let $M_4 \geq |f^{(4)}(x)|$ for all $x \in [a, b]$. Then*

$$\varepsilon(S_n) \leq \frac{(b-a)^5}{180} \cdot M_4 \cdot \frac{1}{n^4}$$

The proof is based on two lemmas. The first finds a cubic polynomial interpolating f. We fit a cubic because Simpson's rule gives the exact value of the integral of a cubic polynomial.

Lemma 4.5 *Suppose f has a continuous derivative on the interval $[a, b]$. Let $m = (a + b)/2$. Then there is a cubic polynomial p such that*

$$f(a) = p(a), f(m) = p(m), f(b) = p(b), \text{ and } f'(m) = p'(m)$$

Proof: An easy calculation. ∎

The second lemma will give us the error estimate.

Lemma 4.6 *Suppose f has four continuous derivatives on $[a, a+2h]$ for some $h > 0$. Let $p(x)$ be the cubic polynomial of Lemma 4.5. Define*

$$\omega(x) = (x - a)(x - a - h)^2(x - a - 2h)$$

Then for every $x \in [a, a + 2h]$ there is a $c \in (a, a + 2h)$, depending on x, such that

$$f(x) - p(x) = \frac{f^{(4)}(c)}{4!} \omega(x)$$

Proof: We'll repeated apply Rolle's theorem to the auxillary function

$$\phi(t) = \omega(x) [f(t) - g(t)] - \omega(t) [f(x) - g(x)]$$

and its derivatives to find a point c where $\phi^{(4)}(c) = 0$. Then

$$0 = \phi^{(4)}(c) = \omega(x) \left[f^{(4)}(c) - g^{(4)}(c) \right] - \omega^{(4)}(c) [f(x) - g(x)]$$
$$= \omega(x) f^{(4)}(c) - 4! \, [f(x) - g(x)]$$

Solving this equation for $f(x) - g(x)$ completes the proof. ∎

Now we return to the proof of the error bound for Simpson's rule.

Proof: Let h be the common distance $x_{k+1} - x_k$. Construct the cubic polynomial p interpolating f given by Lemma 4.5. Since Simpson's rule is exact for cubics, we have

$$\int_{x_k}^{x_{k+2}} p(x)\,dx = \frac{h}{3}\left(p(x_k) + 4p(x_{k+1}) + p(x_{k+2})\right)$$

$$= \frac{h}{3}\left(f(x_k) + 4f(x_{k+1}) + f(x_{k+2})\right)$$

So the error is

$$e = \int_{x_k}^{x_{k+2}} f(x)\,dx - \frac{h}{3}\left(f(x_k) + 4f(x_{k+1}) + f(x_{k+2})\right)$$

$$= \int_{x_k}^{x_{k+2}} f(x)\,dx - \int_{x_k}^{x_{k+2}} p(x)\,dx$$

$$= \int_{x_k}^{x_{k+2}} \left(f(x) - p(x)\right)\,dx$$

Replace $f(x) - p(x)$ according to Lemma 4.6, and take absolute values to see

$$|e| \leq \int_{x_k}^{x_{k+2}} \left| \frac{f^{(4)}(c)}{4!} \omega(x) \right|\,dx$$

$$\leq \frac{M_4}{4!} \int_{x_k}^{x_{k+2}} |\omega(x)|\,dx = \frac{M_4}{4!}\frac{4}{15}h^5$$

We sum the errors over the $n/2$ pairs of panels to have

$$\varepsilon(S_n) \leq \frac{M_4}{4!}\frac{4}{15}h^5\frac{n}{2} = \frac{(b-a)^5}{180}\cdot M_4\cdot\frac{1}{n^4}$$

recalling that $h = (b-a)/n$. Thus the result holds.   ∎

Another way to derive Simpson's rule is to use the endpoints and midpoint of a panel for the three values to fit a parabola. The advantage is that we don't require an even number of panels. Other standard methods used to derive the error bound include using Taylor polynomials, as we did for the midpoint rule, and using error formulas for Lagrange interpolating polynomials.

Table 4.4 compares all four methods.

Table 4.4   Result of Doubling the Number of Panels

Method	Doubling n
Left endpoint rule	accuracy $\times 2$
Trapezoid rule	accuracy $\times 4$
Midpoint rule	accuracy $\times 4$
Simpson's rule	accuracy $\times 16$

Gaussian Quadrature

So far we have chosen partitions with equally spaced points for ease of calculation. A natural question is, "Are there better points to choose than an equipartition?" Let's try.

Since the monotone, bijection $x = (b+a)/2 + u \cdot (b-a)/2$ maps $[a, b]$ to $[-1, 1]$, then

$$\int_a^b f(x)\,dx = \frac{b-a}{2} \int_{-1}^1 f\left(\frac{b+a}{2} + \frac{b-a}{2}\,u\right)\,du$$

Therefore, without loss of generality, we can focus on integrals of the form $\int_{-1}^1 f(x)\,dx$. Looking back at the previous rules, we see they all are weighted averages of function values. That is, they are all variants of

$$\int_{-1}^1 f(x)\,dx \approx \sum_{k=1}^n w_k f(x_k)$$

Our methods so far have chosen $x_k = -1 + (k-1)h$ and treated the w_k's as n unknowns. We found values for w_k by requiring the quadrature to be exact for polynomials of degree $n-1$; i.e., the trapezoid rule integrates linear functions exactly, Simpson's rule integrates quadratics exactly, and so forth.[3] Gauss treated both x_k and w_k as a set of $2n$ unknowns and searched for values that would integrate polynomials of degree $2n-1$ exactly.

■ EXAMPLE 4.1

Let's take $n = 3$. Then the quadratures

$$\int_{-1}^1 x^j\,dx = \sum_{k=1}^3 w_k x_k^j$$

---

[3] Actually, Simpson's rule integrates cubics exactly, but we chose parabolas for approximators to give quadratics exactly.

are exact for $j = 0, \ldots, 5$. We have the system of six nonlinear equations

$$\begin{cases} w_1 + w_2 + w_3 = 2 \\ w_1 x_1 + w_2 x_2 + w_3 x_3 = 0 \\ w_1 x_1^2 + w_2 x_2^2 + w_3 x_3^2 = \dfrac{2}{3} \\ w_1 x_1^3 + w_2 x_2^3 + w_3 x_3^3 = 0 \\ w_1 x_1^4 + w_2 x_2^4 + w_3 x_3^4 = \dfrac{2}{5} \\ w_1 x_1^5 + w_2 x_2^5 + w_3 x_3^5 = 0 \end{cases}$$

A little help from our favorite computer algebra system gives

$$x_1 = -\sqrt{\frac{3}{5}} \qquad x_2 = 0 \qquad x_3 = \sqrt{\frac{3}{5}}$$

and

$$w_1 = \frac{5}{9} \qquad w_2 = \frac{8}{9} \qquad w_3 = \frac{5}{9}$$

Our quadrature rule for $\int_{-1}^{1} f(x)\, dx$ that is exact for polynomials to degree 5 is

$$\int_{-1}^{1} f(x)\, dx \approx \frac{5}{9} f\left(-\sqrt{\frac{3}{5}}\right) + \frac{8}{9} f(0) + \frac{5}{9} f\left(\sqrt{\frac{3}{5}}\right)$$

Let's test the rule on a randomly generated polynomial. Maple's randpoly(x, degree = 5, coeffs = rand(-9 .. 9)) returned

$$8 x^5 + 5 x^4 - 2 x^3 + 4 x + 2$$

Then

$$\sum_{k=1}^{3} w_k p(x_k) = \frac{5}{9}\left(-\frac{142}{125}\sqrt{15} + \frac{19}{5}\right) + \frac{8}{9} \cdot 2 + \frac{5}{9}\left(\frac{142}{125}\sqrt{15} + \frac{19}{5}\right) = 6$$

and so

$$\int_{-1}^{1} p(x)\, dx = 6$$

Our quadrature worked quite nicely. We were able to calculate the integral exactly with only three function evaluations! ■

There is a well-developed theory of Gaussian quadrature that extends to other types of integrals. Given a reasonable positive function $w(x)$, called a *weight function*, we can approximate integrals of the form $\int_a^b f(x)w(x)\, dx$ with Gauss's technique. Several basic forms are listed in Table 4.5. The theory of orthogonal polynomials tells us to choose the x_k to be the roots of certain special polynomials. Choosing these values for x_k reduces finding the w_k to solving an n-dimensional linear system. Extensive tables of w_k and x_k generated by orthogonal polynomials are given in Abramowitz & Stegun (1965).

Table 4.5 Gaussian Quadratures

Quadrature	Integral Form
Gauss-Legendre	$\displaystyle\int_{-1}^{1} f(x)\,dx$
Gauss-Chebyshev	$\displaystyle\int_{-1}^{1} \frac{f(x)}{\sqrt{1-x^2}}\,dx$
Gauss-Laguerre	$\displaystyle\int_{0}^{\infty} f(x)\,e^{-x}\,dx$
Gauss-Hermite	$\displaystyle\int_{-\infty}^{\infty} f(x)\,e^{-x^2}\,dx$

■ **EXAMPLE 4.2**

Find

$$\int_{-1}^{1} \frac{x^7 - 8x^4 + 2x + 5}{\sqrt{1-x^2}}\,dx$$

Refering to Table 4.5, we choose a Gauss-Chebyshev form. Abramowitz & Stegun (1965, p. 889) tells us that for an order n Gauss-Chebyshev quadrature we should use

$$x_k = \cos\left(\frac{2k-1}{2n}\,\pi\right) \qquad \text{and} \qquad w_k = \frac{\pi}{n}$$

The numerator of our integral is degree 7; so letting $2n - 1 = 7$, we choose $n = 4$. This choice of n gives the integral exactly with only four function evaluations! Let $p(x) = x^7 - 8x^4 + 2x + 5$. Hence

$$\int_{-1}^{1} \frac{x^7 - 8x^4 + 2x + 5}{\sqrt{1-x^2}}\,dx = \sum_{k=1}^{4} \frac{\pi}{4} \cdot p\left(\cos\left(\frac{2k-1}{8}\,\pi\right)\right)$$

which is equal to $0.785398\,(1.59385 + 5.59500 + 4.06186 - 3.25071)$ according to Maple. So

$$\int_{-1}^{1} \frac{x^7 - 8x^4 + 2x + 5}{\sqrt{1-x^2}}\,dx = 6.283185$$

Does this number look familiar? We have calculated the exact value of a nontrivial integral with only four function computations. ■

We close this section by stating a theorem of Markov on the error bounds for Gaussian quadratures. The proof will be left to another course.

Theorem 4.7 (Markov) *Let $w(x)$ be a weight function on $[a, b]$, where a and/or b may be infinite; let x_k be the zeros of the nth-degree orthogonal polynomial associated with w, and let w_k be the Gauss-Jacobi weights. If f has a continuous $2n$-th derivative with $M_{2n} \geq |f^{(2n)}(x)|$ for all $x \in [a, b]$, then the error of the Gauss-Jacobi quadrature G_n is*

$$\varepsilon(G_n) = \left| \int_a^b f(x)\, dx - \sum_{k=1}^n w_k f(x_k) \right| \leq c \cdot \frac{M_{2n}}{(2n)!}$$

where c is a positive constant depending on w and $[a, b]$.

For a brief, but thorough, overview of the concepts and uses of numerical quadrature methods, read Zwillinger (1992, Chapters V and VI).

4.3 FOURIER SERIES

In December of 1807, Joseph Fourier presented his manuscript "Théorie du mouvement de la chaleur dans les corps solides" ("On the Propagation of Heat in Solid Bodies") to the Académie des Sciences. Fourier introduced what we now call Fourier series in his paper. This manuscript was controversial, initially rejected, but proved to be amazingly influential. A second version of the manuscript was presented to the Académie in 1811 and subsequently published in 1822 as *Théorie Analytique de la Chaleur (The Analytical Theory of Heat)*.[4] [See Fourier (1822, 2003).] The second version was awarded the Académie's grand prize. At the time, Fourier was in Grenoble as Prefect, or regional governor, of Isère. Napoleon had appointed Fourier to the prefecture in 1802. Fourier would have preferred to stay in his position as Professor of Analysis at l'École Polytechnique in Paris—it was difficult to refuse a request from Napoleon. A brief biography of Fourier's fascinating life is given in O'Connor & Robertson (2008).

One measure of the impact of Fourier's work is that it appears in standard books at all levels today. In calculus: Thomas (1968, Section 18.8), Ostebee & Zorn (2002, Section 9.3), Hughes-Hallett et al. (2009, Section 10.5), Stewart (2009, p. 503 Exercise 70); in real analysis: Rudin (1976, p. 185–192), Apostol (1969, Chapter 11), Davidson & Donsig (2002, Chapter 13). The main theme of Bressoud's excellent set *A Radical Approach to Real Analysis* (2005) and *A Radical Approach to Lebesgue's Theory of Integration* (2008) is the deep and overarching influence of Fourier series on the development of analysis. Bhatia's *Fourier Series* (2005) begins with a brief history of Fourier series from d'Alembert's 1747 "wave equation" to Carleson's 1966 "convergence almost everywhere" results.

In 1755, Daniel Bernoulli, while working on d'Alembert's wave equation, came to believe that any function could be represented by a series like $\sum a_k \sin(k\pi x)$. Euler objected pointing out that Bernoulli's sum had only odd functions. (Remember, the

---

[4]Fourier introduced the notation $\int_a^b$ for definite integrals in his 1822 book, building on the $\int$ of Leibniz.

concepts of "function" and "convergence" were not yet fully developed in the 1700s.) Fourier believed any continuous function could be expanded in a sum of the form

$$\sum_{k=0}^{\infty} a_k \cos(kx) + b_k \sin(kx)$$

He was surprisingly close to being right. Dirichlet published results on Fourier series in 1828 showing that the series converges uniformly to f on intervals where f is continuous and converges to the average value

$$y_a = \frac{f(x_0+) + f(x_0-)}{2} = \frac{1}{2}\left(\lim_{x \to x_0+} f(x) + \lim_{x \to x_0-} f(x)\right)$$

if f has a jump discontinuity at x_0.

Calculating the Fourier Coefficients

We need three simple formulas to help us compute the Fourier coefficients.

Theorem 4.8 *Suppose m and n are integers; then*

1. $\displaystyle\int_{-\pi}^{\pi} \cos(mx)\cos(nx)\,dx = \begin{cases} 0 & m \neq n \\ \pi & m = n \neq 0 \\ 2\pi & m = n = 0 \end{cases}$

2. $\displaystyle\int_{-\pi}^{\pi} \cos(mx)\sin(nx)\,dx = 0$

3. $\displaystyle\int_{-\pi}^{\pi} \sin(mx)\sin(nx)\,dx = \begin{cases} 0 & m \neq n \\ \pi & m = n \neq 0 \\ 0 & m = n = 0 \end{cases}$

Proof: We will show computations as a sketch of the proof.

1. Let m and n be distinct integers. Then

$$\int_{-\pi}^{\pi} \cos(mx)\cos(nx)\,dx = \int_{-\pi}^{\pi} \frac{\cos((m+n)x) + \cos((m-n)x)}{2}\,dx$$

$$= \frac{\sin((m+n)x)}{m-n} + \frac{\sin((m-n)x)}{m+n}\Bigg|_{x=-\pi}^{x=\pi}$$

$$= 0$$

2. Let m and n be distinct integers. Then

$$\int_{-\pi}^{\pi} \cos(mx)\sin(nx)\,dx = \int_{-\pi}^{\pi} \frac{\sin((m+n)x) - \sin((m-n)x)}{2}\,dx$$

$$= \frac{\cos((m-n)x)}{m-n} - \frac{\cos((m+n)x)}{m+n} \Big|_{x=-\pi}^{x=\pi}$$

$$= 0$$

3. Let m and n be distinct integers. Then

$$\int_{-\pi}^{\pi} \sin(mx)\sin(nx)\,dx = \int_{-\pi}^{\pi} \frac{\cos((m-n)x) - \cos((m+n)x)}{2}\,dx$$

$$= \frac{\sin((m-n)x)}{m-n} - \frac{\sin((m+n)x)}{m+n} \Big|_{x=-\pi}^{x=\pi}$$

$$= 0$$

Do the cases where $m = n \neq 0$ and $m = n = 0$ to finish the proof. ■

These three trigonometric integral identities are the main tools needed to develop formulas for the Fourier coefficients. The symmetry of the products shown in Figures 4.9 and 4.10 causes the integrals to be zero. The sum of the areas above and below the x-axis cancel.

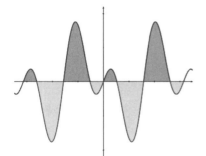

Figure 4.9   Graph of $y = \cos(3x)\sin(x)$ **Figure 4.10**   Graph of $y = \sin(4x)\sin(5x)$

A set of functions $\mathcal{S}$ such that the integral of the product of any two distinct functions from $\mathcal{S}$ is zero, that is,

$$\int_a^b f_1(x)f_2(x)\,dx = 0$$

whenever $f_1, f_2 \in \mathcal{S}$ with $f_1 \neq f_2$, is called an *orthogonal system*. Theorem 4.8 shows that

$$\mathcal{S} = \{1, \cos(kx), \sin(kx) \mid k = 1, 2, \dots\}$$

forms an orthogonal system. *How does this concept relate to "basis" in linear algebra?* We take advantage of orthogonality to calculate the coefficients in a Fourier series.

Suppose that f can be written as the Fourier series

$$f(x) = \frac{a_0}{2} + \sum_{k=1}^{\infty} a_k \cos(kx) + b_k \sin(kx)$$

(We're using $a_0/2$ to simplify the formulas.) Multiply both sides by $\cos(mx)$. Then

$$\cos(mx)f(x) = \frac{a_0}{2}\cos(mx) + \sum_{k=1}^{\infty} a_k \cos(mx)\cos(kx) + b_k \cos(mx)\sin(kx)$$

Integrate all terms for x from $-\pi$ to π. *What convergence properties do we need to integrate the series term by term?* Applying Theorem 4.8 to each integral leaves only the cosine term with $k = m$, all other terms must be zero;

$$\int_{-\pi}^{\pi} \cos(mx)f(x)\,dx = a_m \int_{-\pi}^{\pi} \cos(mx)\cos(mx)\,dx$$

for $m \neq 0$, and

$$\int_{-\pi}^{\pi} f(x)\,dx = \frac{a_0}{2} \int_{-\pi}^{\pi} dx$$

for $m = 0$. Apply Theorem 4.8 again, then solve these expressions for a_m and a_0, respectively, to see

$$a_m = \frac{1}{\pi} \int_{-\pi}^{\pi} \cos(mx)f(x)\,dx$$

and

$$a_0 = \frac{1}{\pi} \int_{-\pi}^{\pi} f(x)\,dx$$

We used $a_0/2$ in the Fourier series so the constants in these formulas would match. Some texts do, some don't.

Carry out the same procedure with $\sin(mx)$ to have

$$b_m = \frac{1}{\pi} \int_{-\pi}^{\pi} f(x)\sin(mx)\,dx$$

■ **EXAMPLE 4.3**

Find the Fourier series for the *square wave* with period π,

$$f(x) = \begin{cases} +1 & -\pi \leq x < -\pi/2 \\ -1 & -\pi/2 \leq x < 0 \\ +1 & 0 \leq x < \pi/2 \\ -1 & \pi/2 \leq x < \pi \end{cases}$$

[An alternate definition for f is $f(x) = \text{sign}(\sin(2x))$.] *Graph $f(x)$!*

Calculate the Fourier coefficients. First observe that f is an odd function. Therefore $f(x)\cos(kx)$ is odd for all k, and so $a_k = 0$. The integrals for b_k are

$$b_k = \frac{1}{\pi}\int_{-\pi}^{\pi} f(x)\sin(kx)\,dx$$

$$= \frac{1}{\pi}\int_{-\pi}^{-\pi/2}\sin(kx)\,dx + \frac{1}{\pi}\int_{-\pi/2}^{0} -\sin(kx)\,dx$$

$$+ \frac{1}{\pi}\int_{0}^{\pi/2}\sin(kx)\,dx + \frac{1}{\pi}\int_{\pi/2}^{\pi} -\sin(kx)\,dx$$

$$= \frac{2}{k\pi}\left(1 + \cos(k\pi) - 2\cos\left(\frac{k\pi}{2}\right)\right)$$

$$= \frac{2}{\pi} \cdot \begin{cases} \dfrac{2}{2k-1} & k \bmod 4 = 2 \\ 0 & \text{otherwise} \end{cases}$$

Verify these calculations!

Hence, the partial sum S_N of the Fourier series for f is given by

$$S_N(x) = \frac{4}{\pi}\sum_{k=1}^{\lfloor (N+2)/4\rfloor}\frac{\sin(2(2k-1)x)}{2k-1}$$

Figures 4.11 to 4.14 show f along with selected partial sums.   ∎

Research the Gibbs phenomenon!

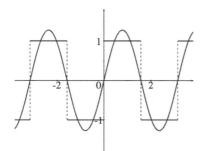

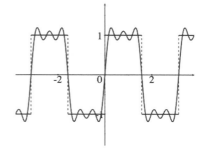

Figure 4.11   Plot of f and S_2 **Figure 4.12**   Plot of f and S_{10}

The Fourier Series of a Periodic Function

Collect the Fourier formulas for reference.

$$\hat{f}(x) = \frac{a_0}{2} + \sum_{k=1}^{\infty} a_k\cos(kx) + b_k\sin(kx)$$

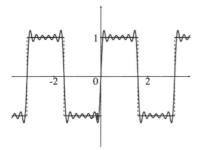

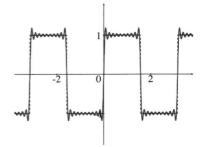

Figure 4.13 Plot of f and S_{22} **Figure 4.14** Plot of f and S_{38}

where

$$a_k = \frac{1}{\pi} \int_{-\pi}^{\pi} f(x) \cos(kx)\, dx \qquad k \geq 0$$

$$b_k = \frac{1}{\pi} \int_{-\pi}^{\pi} f(x) \sin(kx)\, dx \qquad k \geq 1$$

Table 4.6 lists several standard Fourier series.

Table 4.6   Standard Fourier Series

Function on $[-\pi, \pi]$	Fourier Series
$f(x) = x$	$2 \displaystyle\sum_{k=1}^{\infty} (-1)^{k+1} \frac{\sin(kx)}{k}$
$f(x) = \|x\|$	$\dfrac{\pi}{2} - \dfrac{4}{\pi} \displaystyle\sum_{k=1}^{\infty} \frac{\cos((2k-1)x)}{(2k-1)^2}$
$f(x) = x^2$	$\dfrac{\pi^2}{3} + 4 \displaystyle\sum_{k=1}^{\infty} (-1)^k \frac{\cos(kx)}{k^2}$
$f(x) = \sin^2(x)$	$\dfrac{1}{2} - \dfrac{1}{2} \cos(2x)$
$f(x) = \|\sin(x)\|$	$\dfrac{\pi}{2} - \dfrac{4}{\pi} \displaystyle\sum_{k=1}^{\infty} \frac{\cos(2kx)}{4k^2 - 1}$
$f(x) = \begin{cases} \dfrac{2\pi - 4\|x\|}{\pi^2} & \|x\| < \pi/2 \\ 0 & \text{otherwise} \end{cases}$	$\dfrac{1}{2\pi} + \dfrac{8}{\pi^3} \displaystyle\sum_{k=1}^{\infty} \frac{1 - \cos(k\pi/2)}{k^2} \cos(kx)$

Remember that the functions in Table 4.6 are periodically extended beyond $[-\pi, \pi]$. Graphs of the first and last functions in the table are shown in Figures 4.15 and 4.16, respectively.

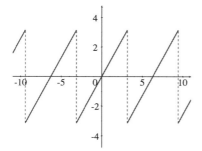

Figure 4.15   Periodic Sawtooth Wave

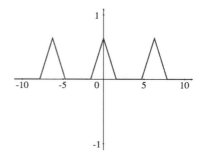

Figure 4.16   Periodic Triangle Pulse

There is an integral representation for Fourier series that is used to prove that the series converges uniformly on intervals where f is continuous. This representation uses a special function called the *Dirichlet kernel*,

$$D_n(x) = \frac{\sin\left(\left(n + \frac{1}{2}\right)x\right)}{2\sin\left(\frac{1}{2}x\right)}$$

The partial sums S_n can then be expressed as

$$S_n(x) = \frac{1}{\pi}\int_{-\pi}^{\pi} f(x+u)\frac{\sin\left(\left(n+\frac{1}{2}\right)u\right)}{2\sin\left(\frac{1}{2}u\right)}\,du$$

Using this expansion would let us prove Dirichlet's theorem of 1828.

Theorem 4.9 (Dirichlet) *Let f be 2π-periodic and piecewise smooth, and let S_n be the nth partial sum of the Fourier series of f. Then*

$$\lim_{n\to\infty} S_n(x) = \frac{f(x+) + f(x-)}{2}$$

Moreover, if f is continuous at x, then $\lim S_n(x) = f(x)$.

For the details of this proof and further development of the theory, see Jackson (1941), Körner (1989), or Bhatia (2005).

The Spectrum of a Fourier Series

Fourier series are widely used in applications ranging from image analysis and audio analysis to earthquake geology and structural engineering to economics and biology. Special versions of Fourier series called *fast Fourier transforms* can even be used to multiply very large numbers efficiently. Often in applications, the focus of interest isn't the Fourier series itself, but its coefficients.

Define the *spectrum* of a Fourier series to be the amplitude of the harmonics $a_k\cos(kx) + b_k\sin(kx)$; that is, the spectrum is a function from $\mathbb{N}$ to $\mathbb{R}$ given by $\phi(k) = \sqrt{a_k^2 + b_k^2}$. A physical interpretation is the spectrum indicates the amount of energy of a particular harmonic. A plot of the spectrum is called the *line spectrum*.

■ EXAMPLE 4.4

Plot the line spectrum for the square wave $f(x) = \text{sign}(\sin(2x))$, the function
of Example 4.3.

We have already determined that

$$a_k = 0 \quad \text{and} \quad b_k = \frac{2}{\pi} \cdot \begin{cases} \dfrac{2}{2k-1} & k \bmod 4 = 2 \\ 0 & \text{otherwise} \end{cases}$$

Thus $\phi(k) = b_k$. Graphing vertical lines from $(k, 0)$ to (k, b_k) produces the line
spectrum shown in Figure 4.17.   ■

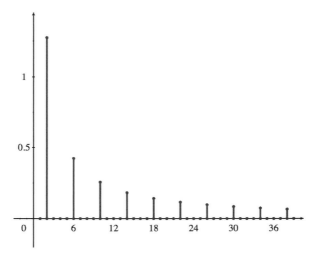

Figure 4.17   The Line Spectrum of a Square Wave

A very interesting application of Fourier series, especially of line spectrum graphs, is
comparing musical instruments. Figure 4.18 shows the line spectrum for a flute, an
oboe, and a violin, each playing A_4, the A above middle C. A line spectrum graph
allows us to compare instruments or assists with creating electronically generated
music. Line spectrum graphs are used widely; other applications include measuring
flow and pressure in arteries, measuring biological morphology, and so forth. A
delightful description of applying Fourier analysis to the opening chord from the
Beatle's *A Hard Day's Night* appears in Brown (2004).

While not being overly modest, Fourier was actually understating the importance
and applicability of his work when he stated in the second paragraph of the "Preliminary
Discourse" in his *The Analytical Theory of Heat* that

> *The theory of heat will hereafter form one of the most important branches of
> general physics.*

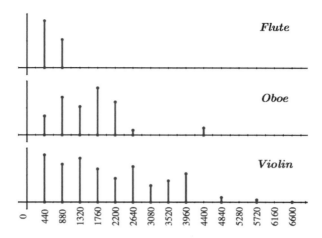

Figure 4.18 Line Spectra of Flute, Oboe, and Violin Playing A_4

4.4 SPECIAL FUNCTIONS—THE GAMMA FUNCTION

Euler believed that a function must be able to be written as a single, simple formula. For him,

$$f(x) = |x| = \begin{cases} x & x \geq 0 \\ -x & x < 0 \end{cases}$$

was *not* a function. (Weierstrass was the first to use $|\cdot|$ for absolute value, introducing the symbol in an 1841 essay.) Lobachevsky and Dirichlet were the first to give the modern definition of a correspondence between two sets. With our broad concept, most functions do not have simple expressions. Many non-simple functions occur in such significant ways that they have been given names; the collection of these is called *special functions*. Examples of important special functions include the *Airy function*

$$\mathrm{Ai}(x) = \frac{1}{\pi} \int_0^\infty \cos\left(\frac{1}{3}t^3 + tx\right) dt$$

the *error function*

$$\mathrm{erf}(x) = \frac{1}{\sqrt{\pi}} \int_0^x e^{-t^2} dt$$

the *basic hypergeometric function*

$$_2F_1(a, b; c; x) = \sum_{n=0}^\infty \frac{(a)_n (b)_n}{(c)_n} \frac{x^n}{n!}$$

where $(r)_n = r(r+1)(r+2) \cdots (r+n-1) = (r+n-1)!/(r-1)!$ and Riemann's *zeta function*

$$\zeta(x) = \sum_{n=1}^{\infty} \frac{1}{n^x}$$

We will focus on the gamma function—a generalization of factorial. This function is one of the simplest special functions to define and yet one of the most useful and widely applied. Definitions come in many varieties: some special functions are defined by differential equations, others by "unsolvable equations." Bessel functions are the solutions of $x^2 y'' + xy' + (x^2 - \alpha^2) = 0$; Lambert's w-function is the solution of the equation $x = w(x)e^{w(x)}$.

The Gamma Function

Euler introduced the gamma function as an infinite product in 1729, then gave an equivalent integral in 1730. We'll use the integral form as our definition.

$$\Gamma(x) = \int_0^{\infty} t^{x-1}e^{-t}\,dt.$$

Legendre is credited with choosing the symbol Γ; Euler had used $C(x)$.

Let's begin by computing $\Gamma(1)$.

$$\Gamma(1) = \int_0^{\infty} t^{1-1}e^{-t}\,dt = \int_0^{\infty} e^{-t}\,dt$$
$$= e^0 - \lim_{x \to \infty} e^{-x} = 1$$

Now let's integrate $\Gamma(x+1)$ by parts. Assume x is not a negative integer; then

$$\Gamma(x+1) = \int_0^{\infty} t^x e^{-t}\,dt$$
$$= t^x e^{-t}\Big|_{t=0}^{t \to \infty} + x\int_0^{\infty} t^{x-1}e^{-t}\,dt$$
$$= x\int_0^{\infty} t^{x-1}e^{-t}\,dt = x\,\Gamma(x)$$

We have the recurrence relation $\Gamma(x+1) = x\,\Gamma(x)$. *Where did we use the assumption that x was not a negative integer?* Let $n \in \mathbb{N}$, and iterate the recurrence to obtain

$$\Gamma(n+1) = n\,\Gamma(n)$$
$$= n(n-1)\Gamma(n-1)$$
$$= n(n-1)(n-2)\Gamma(n-2)$$
$$\vdots$$
$$= n(n-1)(n-2)\cdots 3 \cdot 2 \cdot \Gamma(1)$$
$$= n!$$

The formula $\Gamma(n+1) = n!$ is why we say Γ extends the factorial function. Figure 4.19 shows $\Gamma(x)$ graphed in the window $[-7, 5] \times [-7, 7]$. Note the poles at the negative integers. Observe that $\Gamma(1) = 1$ gives another reason for setting $0! = 1$.

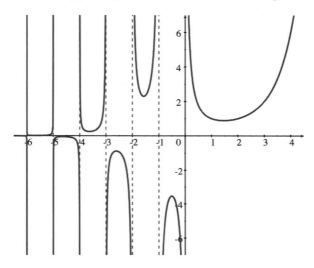

Figure 4.19 Graph of $\Gamma(x)$

Let's consider a general property of the gamma function called the *reflection formula* that is due to Euler. The easiest proof uses Weierstrass's product formula

$$\Gamma(x) = \frac{e^{-\gamma x}}{x} \prod_{n=1}^{\infty} \left(1 + \frac{x}{n}\right)^{-1} e^{x/n}$$

where γ is *Euler's constant* ≈ 0.57722. Euler's constant is the difference between the nth partial sum of the harmonic series and $\ln(n)$. The harmonic series is related to the gamma function which is related to factorials! Weierstrass's formula is based on Gauss's limit

$$\Gamma(x) = \lim_{n \to \infty} \frac{n^x \, n!}{x(x+1)\ldots(x+n)}$$

See the classic text Rainville (1971) for a derivation.

Theorem 4.10 (Euler's Reflection Formula) *Suppose that $x \in \mathbb{R}$ is not an integer. Then*

$$\Gamma(x) \cdot \Gamma(1 - x) = \frac{\pi}{\sin(\pi x)}$$

Proof: Invert Weierstrass's product formula to see

$$\frac{1}{\Gamma(x)} = x e^{\gamma x} \prod_{n=1}^{\infty} \left(\left(1 + \frac{x}{n}\right) e^{-x/n}\right)$$

Multiply $1/\Gamma(x)$ by $1/\Gamma(-x)$ to get

$$\frac{1}{\Gamma(x)} \cdot \frac{1}{\Gamma(-x)} = -x^2 e^{\gamma x} e^{-\gamma x} \prod_{n=1}^{\infty} \left(\left(1 + \frac{x}{n}\right) e^{-x/n} \right) \left(\left(1 - \frac{x}{n}\right) e^{x/n} \right)$$

$$= -x^2 \prod_{n=1}^{\infty} \left(1 - \frac{x^2}{n^2}\right)$$

The recurrence relation gives us $\Gamma(1 - x) = -x\Gamma(-x)$, so

$$\frac{1}{\Gamma(x)} \cdot \frac{1}{\Gamma(1 - x)} = x \prod_{n=1}^{\infty} \left(1 - \frac{x^2}{n^2}\right)$$

The infinite product on the right side is equal to $\sin(\pi x)/\pi$ (see p. 123). Hence

$$\frac{1}{\Gamma(x)} \cdot \frac{1}{\Gamma(1 - x)} = \frac{\sin(\pi x)}{\pi}$$

and the theorem follows. ∎

Two formulas, important in probability, are an immediate consequence of Euler's reflection formula.

Corollary 4.11

$$\Gamma\left(\frac{1}{2}\right) = \sqrt{\pi} \quad and \quad \Gamma\left(\frac{3}{2}\right) = \frac{\sqrt{\pi}}{2}$$

Write the formulas from Corollary 4.11 as integrals.
   There is a large collection of properties of the gamma function. To go into more depth would require complex variables, so we will look at an important approximation.

Stirling's Formula

Stirling's original formula, which appeared in his 1726 book *Methodus Differentialis*, applied only to factorials; Laplace extended the result to the gamma function in 1774.

Theorem 4.12 (Stirling's Formula) *For large x,*

$$\Gamma(x + 1) \approx \left(\frac{x}{e}\right)^x \sqrt{2\pi x}$$

In particular, for large integers n, it follows that $n! \approx (n/e)^n \sqrt{2\pi n}$.

Following Rudin (1976, p. 200, Exercise 20), we'll consider a simplified argument that "almost proves" Stirling's formula.

Proof: Define the two functions f and g as follows. For each $m \in \mathbb{N}$, if $m \le x < m + 1$, set

$$f(x) = (m + 1 - x)\ln(m) + (x - m)\ln(m + 1)$$

and if $m - 1/2 \le x < m + 1/2$, then let

$$g(x) = \frac{x}{m} - 1 + \ln(m).$$

If $x \ge 1$, then $f(x) \le \ln(x)$ because f is the secant line from $(m, \ln(m))$ to $(m + 1, \ln(m + 1))$. Since g is the tangent line at $(m, \ln(m))$ and $\ln(x)$ is concave down, we also have $\ln(x) \le g(x)$.

Integrate $\ln(x)$ from 1 to n. Then

$$\int_1^n \ln(x)\, dx = n \ln(n) - n + 1$$

Now integrate $f(x)$ from m to $m + 1$ to have

$$\int_m^{m+1} f(x)\, dx = \frac{1}{2} \ln(m) + \frac{1}{2} \ln(m + 1)$$

Then

$$\int_1^n f(x)\, dx = \sum_{m=1}^{n-1} \frac{1}{2} \ln(m) + \frac{1}{2} \ln(m + 1)$$

$$= \frac{1}{2} \ln((n - 1)!) + \frac{1}{2} \ln(n!) = \ln(n!) - \frac{1}{2} \ln(n)$$

Since $f(x) \le \ln(x)$, it follows that

$$\int_1^n f(x)\, dx \le \int_1^n \ln(x)\, dx$$

$$\ln(n!) - \frac{1}{2} \ln(n) \le n \ln(n) - n + 1$$

So

$$\ln(n!) - \left(n + \frac{1}{2}\right) \ln(n) + n \le 1$$

Now integrate g from $m - 1/2$ to $m + 1/2$ to have

$$\int_{m-1/2}^{m+1/2} g(x)\, dx = \ln(m)$$

Then

$$\int_1^n g(x)\, dx = \int_1^{3/2} g(x)\, dx + \sum_{m=2}^{n-1} \int_{m-1/2}^{m+1/2} g(x)\, dx + \int_{n-1/2}^n g(x)\, dx$$

$$= \frac{1}{8} + \sum_{m=2}^{n-1} \ln(m) + \left(\frac{1}{2} \ln(n) - \frac{1}{8n}\right)$$

$$= \ln(n!) - \frac{1}{2} \ln(n) - \frac{1}{8n} + \frac{1}{8}$$

Since $\ln(x) \leq g(x)$, it follows that

$$\int_1^n \ln(x)\, dx \leq \int_1^n g(x)\, dx$$

$$n\ln(n) - n + 1 \leq \ln(n!) - \frac{1}{2}\ln(n) - \frac{1}{8n} + \frac{1}{8}$$

$$< \ln(n!) - \frac{1}{2}\ln(n) + \frac{1}{8}$$

Therefore

$$\frac{7}{8} \leq \ln(n!) - \left(n + \frac{1}{2}\right)\ln(n) + n$$

Combine the upper and lower bounds

$$\frac{7}{8} \leq \ln(n!) - \left(n + \frac{1}{2}\right)\ln(n) + n \leq 1$$

Then exponentiate

$$e^{7/8} \leq \frac{n!}{(n/e)^n \sqrt{n}} \leq e$$

And the theorem is "almost proven."   ∎

How do the constants $e^{7/8}$, $\sqrt{2\pi}$, and e relate?
Using more sophisticated techniques gives more precise versions of Stirling's formula, such as

$$\Gamma(n+1) = \sqrt{2\pi n}\left(\frac{n}{e}\right)^n\left(1 + \frac{1}{12n} + \frac{1}{288n^2} - \frac{139}{5140n^3} + O\left(\frac{1}{n^4}\right)\right)$$

A similar expression

$$\left[8x^3 + 4x^2 + x + \frac{1}{100}\right]^{1/6} < \frac{\Gamma(x+1)}{(x/e)^x \sqrt{\pi}} < \left[8x^3 + 4x^2 + x + \frac{1}{30}\right]^{1/6}$$

for $x > 0$ is known as *Ramanujan's double inequality*. Read Diaconis & Freedman (1986) for a brief history of Stirling's formula and a version of Laplace's proof. See Davis (1959) for a very good history of Euler's gamma function.

4.5 CALCULUS WITHOUT LIMITS: DIFFERENTIAL ALGEBRA

While numerical integration, or quadrature, is one of the oldest calculus topics, the study of finding antiderivatives algorithmically, especially with technology, is one of the youngest. There are only a handful of basic derivative formulas needed to deal with any function a student of calculus is likely to encounter. However, current calculus books' endpapers often show over a hundred integral formulas [cf. Stewart (2009)]; mathematics handbooks can have 1000s. Most early results were ad hoc, dealing with specific cases. The study of general algorithmic integration began with Laplace's theorem of 1812 stating that rational functions always had elementary antiderivatives. In 1834, Liouville extended Laplace's work to show that if the antiderivative of an algebraic function is elementary, then the antiderivative is the sum of an algebraic function and, possibly, a finite number of logarithms of algebraic functions. Liouville's work led to the name "integration in finite terms."

Slagle's 1961 program calculating the integrals of "freshman calculus at Harvard" began the implementation of technology in algorithmic integration. In 1968, Risch presented his doctoral dissertation on antidifferentiation and published the results in 1969 and 1970. Risch gave a decision procedure, now called the *Risch algorithm*, that either provided an elementary antiderivative for a function or proved that one did not exist. Risch's algorithm depends on the algebraic structure of a function; his technique is purely algebraic and doesn't use limits. Maple and Mathematica have partial implementations of the Risch algorithm.

The Real Differential Algebra

A *differential algebra* provides an alternate method of defining derivatives, and subsequently antiderivatives, without using limits. We take several theorems from elementary calculus as assumed properties of an operator D.

Definition 4.1 *An operator D on $\mathbb{R}(x)$ is a* derivation *if and only if for any $c \in \mathbb{R}$ and any α and $\beta \in \mathbb{R}(x)$ we have*

1. $D(x) = 1$,

2. $D(c\alpha) = c\,D(\alpha)$ *for any constant c,*

3. $D(\alpha + \beta) = D(\alpha) + D(\beta)$,

4. $D(\alpha \cdot \beta) = D(\alpha) \cdot \beta + \alpha \cdot D(\beta)$ *(called the* Leibniz rule*).*

Pairing $\mathbb{R}(x)$ with D gives us a *differential field*, and, since we can multiply elements of $\mathbb{R}(x)$ thinking of them as vectors, $\mathbb{R}(x)$ with D forms a *differential algebra*.

Several properties follow directly from the definition.

Theorem 4.13 *Let $c \in \mathbb{R}$ and $n \in \mathbb{N}$. Then*

1. $D(1) = 0$

2. $D(c) = 0$

3. $D(x^n) = n\,x^{n-1}$

4. $D(x^{-n}) = -n\,x^{-n-1}$

Proof:

1. Apply the Leibinz rule to $1 = 1 \cdot 1$ to see $D(1) = D(1 \cdot 1) = D(1) \cdot 1 + 1 \cdot D(1) = 2D(1)$. Therefore $D(1) = 0$.

2. Apply the linearity of D to $c = c \cdot 1$ to have $D(c) = D(c \cdot 1) = c \cdot D(1) = c \cdot 0 = 0$.

3. For $n = 1$, we have $D(x) = 1$ by definition. Suppose the result holds for some $k \in \mathbb{N}$. Then $D(x^{k+1}) = D(x^k \cdot x) = D(x^k) \cdot x + x^k \cdot D(x)$. By hypothesis, this expression becomes $D(x^{k+1}) = kx^{k-1} \cdot x + x^k = (k+1)x^k$. The result holds by induction on k.

4. Since $1 = x^{-n} \cdot x^n$, then $0 = D(x^{-n} \cdot x^n) = D(x^{-n}) \cdot x^n + x^{-n} \cdot D(x^n) = D(x^{-n}) \cdot x^n + x^{-n} \cdot nx^{n-1}$. Solving this expression for $D(x^{-n})$ yields the result.

■

We now have all of polynomial calculus.

Typically, there is no *chain rule* in differential algebra. Since our goal is *anti*differentiation, we will add the rule.

Definition 4.2 (Chain Rule) *For α and $\beta \in \mathbb{R}(x)$, define*

$$D(\alpha \circ \beta) = (D(\alpha) \circ \beta) \cdot D(\beta)$$

With a chain rule, we can add differentiation of roots and rational functions.

Theorem 4.14 *Let p and q be integers and α and β be compositions of functions in $\mathbb{R}(x)$. Then*

1. $D(x^{1/q}) = \dfrac{1}{q} x^{1/q-1}$

2. $D(x^{p/q}) = \dfrac{p}{q} x^{p/q-1}$

3. $D\left(\dfrac{\alpha}{\beta}\right) = \dfrac{D(\alpha) \cdot \beta - \alpha \cdot D(\beta)}{\beta^2}$

Proof:

1. Since $\left(x^{1/q}\right)^q = x$, we have $q \left(x^{1/q}\right)^{q-1} D\left(x^{1/q}\right) = 1$.

2. *Exercise!*

3. *Exercise!*

■

With composition, roots, and rational functions, we can handle almost any elementary function. What's missing? A simple function eludes us. While differentiating x^{-1} is trivial, antidifferentiating x^{-1} still presents a problem. We fix this in the "standard mathematical way," by defining a new function! Define the function L by the relation: for any nonzero α,

$$D(L \circ \alpha) = \dfrac{D(\alpha)}{\alpha}$$

Since $D(L(x)) = D(x)/x = 1/x$, we have an antiderivative for $1/x$.

The last missing function is the one for which the relation $\alpha' = \alpha$ holds. We use the same technique. Define the function E by the relation: for any nonzero α,

$$D(E \circ \alpha) = E(\alpha) \cdot D(\alpha)$$

What are the common names of L and E?

We now have the collection of elementary functions over $\mathbb{R}$: any finite algebraic combination of functions from $\mathbb{R}(x)$ or any roots, applications of L or E, or compositions of these. The elementary functions form a rather large class!

Formal Integration

With elementary functions defined, we're ready to work towards Liouville's result.

Definition 4.3 *The* integral *of a function f is any function F such that $D(F) = f$; we write $\int f = F$.*

The definition implies that if F and G are any two integrals of f, then $D(F - G) = 0$, so F and G differ by a constant, where by "constant" we mean anything with a zero derivative.

We must go forward cautiously, however, as standard properties that we proved with limits may now fail. For example, $\int f + g = \int f + \int g$ only when at least two of the three integrals exist: $\int x^x + \ln(x)\, x^x \neq \int x^x + \int \ln(x)\, x^x$.

Theorem 4.15 (Liouville's Principle) *If a function f has an elementary integral, then the integral has the form*

$$\int f = F_0 + \sum_{k=1}^{n} c_k L(F_k)$$

where the c_i are constants and the F_i are functions made from $\mathbb{R}(x)$.

Proof: (Sketch.) For a rational function, we decompose it into a polynomial part f_0 and a proper fraction $p(x)/q(x)$. Integrating f_0 is easy. Expand p/q in partial fractions as a sum of terms $c_k/(x - r_i)^{j_{k,r_i}}$ where the c_i are constants, the r_i are the roots of q, and the $j = 1, 2, \ldots, (multiplicity\ of\ r_i)$. The integrals of the partial fraction terms are either of the form $c_k L(x - r_i)$ or rational functions of x. If any roots are complex, they must occur in conjugate pairs. These paired L terms can be combined to eliminate complex numbers. ∎

There has been a great deal of work on implementing algorithmic integration since the first attempts of Slagle in 1961. Risch's algorithm has been studied and improved, but the underlying mathematics is so complex that even the most powerful computer algebra systems still have only partial implementations. New research is going on as we read.

We'll close this section by showing that $\int e^{x^2}$ has no elementary integral.

Theorem 4.16 *The function $f(x) = e^{x^2}$ has no elementary integral.*

Proof: Suppose that f has an elementary integral F. Then by Liouville's principle and the fact that $D(e^{x^2}) = e^{x^2} \cdot 2x$, we see that $F(x) = r(x) \cdot e^{x^2}$. Therefore

$$F'(x) = r'(x)e^{x^2} + r(x) \cdot 2x\, e^{x^2} = e^{x^2}$$

Dividing by e^{x^2} gives the differential equation

$$r' + 2xr = 1$$

Observe that r cannot be polynomial since the left side has degree greater than zero while the right side has degree zero. So r must be a rational function. Let $r(x) = p(x)/q(x)$ with $\gcd(p(x), q(x)) = 1$. Since q cannot be a constant polynomial (*Why not?*), the fundamental theorem of algebra implies q must a have a root z_0, which may be complex. Then $p(z_0) \neq 0$ because p and q have no common factors. Write

$$r(x) = \frac{h(x)}{(x - z_0)^m}$$

where h is a rational function that is nonzero at z_0 and m is the multiplicity of the root. Substitute this form of r into the differential equation. Then

$$1 = \left(\frac{h(x)}{(x - z_0)^m} \right)' + 2x \frac{h(x)}{(x - z_0)^m}$$

that is,

$$1 = \frac{h'(x)}{(x - z_0)^m} - m \frac{h(x)}{(x - z_0)^{m+1}} + 2x \frac{h(x)}{(x - z_0)^m}$$

Taking the limit of both sides as $x \to z_0$ leads to the contradiction $1 = \infty$. Hence r cannot be a rational function. Thus, by Liouville's principle, e^{x^2} does not have an elementary antiderivative.   ∎

Liouville's principle led to the conclusion that a "simple" elementary function may not have an elementary integral. There are a plethora of examples that are special functions including the *sine integral* $\int \sin(x)/x$, the *Fresnel integrals* $\int \sin(\pi t^2/2)$ and $\int \cos(\pi t^2/2)$, and, of course, the gamma function $\int e^{-t}t^{x-1}$.

To learn more about Liouville theory and integration in finite terms, start with Marchisotto & Zakeri (1994), who present the development of algorithmic integration and give statements of the major theorems. Another "limitless" approach to calculus aimed at simplifying differentiation was given by Carathéodory; read Kuhn (1991) for an introduction.

Summary

We have looked at several special topics that are closely related to elementary calculus, but we have studied them from an advanced viewpoint. We used numeric differentiation and minimizing a function via derivatives to fit a logistic model. We used weighted sums involving roots of polynomials to approximate a definite integral. We expanded a function in a series based on trigonometric functions rather than powers of x. We studied a function defined by an integral, instead of an algebraic formula. We ended by looking at differentiation and integration formally, without using limits—avoiding one of the major stumbling blocks in the logical development of calculus. Now, it's time to reflect on all that we've done, to differentiate the techniques and analyses, to integrate the concepts into our own calculus framework, and to study new directions as real analysis keeps growing.

EXERCISES

Modeling with Logistic Functions— Numeric Derivatives

4.1 Report on a current use of the logistic model based on a paper you find in a research journal.

4.2 The *Gompertz population model* is

$$\frac{dp}{dt} = r\, p \ln\left(\frac{m}{p}\right)$$

Write a report comparing the logistic and Gompertz growth models.

4.3 Show that for the logistic function

$$p(t) = \frac{M}{1 + be^{-rt}}$$

a) $\lim_{t \to +\infty} p(t) = M$.
b) $\lim_{t \to -\infty} p(t) = 0$.
c) $p'(t) > 0$ for all t.
d) The only inflection point occurs at $(\ln(b)/r, M/2)$.

4.4 Use the forward difference approximation of the derivative to fit a logistic curve to the population data in Table 4.1. Compare your fit to Pearl, Reed, and Kish's result.

4.5 Data and projections for the world population appear in the following table.

Year	World Population (billions)
1950	2.556
1960	3.039
1970	3.707
1980	4.454
1990	5.279
2000	6.083
2010	6.849*
2020	7.585*
2030	8.247*
2040	8.850*
2050	9.346*
2060	. . .

* Projected value.

Data source: U.S. Census Bureau, International Database. http://www.census.gov/

a) Use a logistic model to check the Census Bureau projections.
b) What is your projection for 2060?
c) What does your model give for the *carrying capacity* of the world?

4.6 Find the census data for your home state from the Census Bureau (www.census.gov). Fit a logistic model, then predict the next census count.

4.7 Data and projections for the population of China appear in the following table.

Year	Population (thousands)
1950	562,580
1960	650,661
1970	820,403
1980	984,736
1990	1,138,895
2000	1,262,474
2010	1,342,783*
2020	. . .

* Projected value.

Data source: U.S. Census Bureau, International Database. http://www.census.gov/

a) Use a logistic model to check the Census Bureau projections.

b) What is your projection for 2020?

c) What does your model give for the *carrying capacity* of China?

4.8 Data and projections for the population of India appear in the following table.

Year	Population (thousands)
1950	369,880
1960	445,857
1970	555,043
1980	687,029
1990	841,655
2000	1,002,708
2010	1,155,011*
2020	. . .

* Projected value.

Data source: U.S. Census Bureau, International Database. http://www.census.gov/

a) Use a logistic model to check the Census Bureau projections.

b) What is your projection for 2020?

c) What does your model give for the *carrying capacity* of India?

4.9 What event of 1918 greatly affected the U.S. and world populations?

4.10 An economic interpretation of the inflection point of a logistic curve calls it the *point of diminishing returns*. Explain this terminology.

4.11 Report on the *Lotka-Volterra population model* for a predator-prey system.

4.12 Report on the *competitive hunter population model* for two species.

Numerical Quadrature

4.13 Report on *Hippocrates' (of Chios) quadrature of the lune.*

4.14 Report on *Archimedes' quadrature of the parabola.*

4.15 Create an annotated set of links to interactive quadrature demonstrations on the internet.

4.16 What do Maple, Mathematica, and TI-Nspire return for the indefinite integral

$$\int \cos(8\sin(x-1)) + 1 \, dx$$

4.17 Analyze the right endpoint rule in the same manner that we did the left.

4.18 Show that the trapezoid rule is the average of the left and right endpoint rules.

4.19 Let $h = (b-a)/n$, $m_k = (x_{k-1} + x_k)/2$, and $y_k = f(x_k)$.

a) Integrate

$$P_k = \int_{x_{k-1}}^{x_k} (t - m_k) f'(t) \, dt$$

by parts to show

$$P_k = \frac{h}{2}(y_{k-1} + y_k) - \int_{x_{k-1}}^{x_k} f(x) \, dx$$

b) Integrate

$$Q_k = \frac{1}{2} \int_{x_{k-1}}^{x_k} \left[\left(\frac{h}{2}\right)^2 - (t - m_i)^2 \right] f''(t) \, dt$$

by parts to show $Q_k = P_k$.

4.20   Let $f(x) = \cos(8\sin(x-1))+1$.
   a) Find an upper bound for $|f'(x)|$
   on $[0, 3]$.
   b) Determine n so that L_n has error
   less than 10^{-5}.
   c) Find an upper bound for $|f''(x)|$
   on $[0, 3]$.
   d) Determine n so that MP_n has
   error less than 10^{-5}.
   e) Find an upper bound for
   $|f^{(4)}(x)|$ on $[0, 3]$.
   f) Determine n so that S_n has error
   less than 10^{-5}.

4.21   The system of equations

$$y_k = a + bx_k + cx_k^2,$$
$$y_{k+1} = a + b(x_k + h) + c(x_k + h)^2,$$
$$y_{k+2} = a + b(x_k + 2h) + c(x_k + 2h)^2$$

defines a parabola $p(x) = a + bx + cx^2$.
   a) Solve the system for the coefficients a, b, and c.
   b) Find

$$\int_{x_k}^{x_k+2h} p(x)\,dx$$

4.22   Use all the methods discussed to
estimate

$$\int_{-5}^{5} e^{-x^2}\,dx$$

and find error estimates for each value
that you calculate.

4.23   Use all the methods discussed to
estimate

$$\int_{0}^{1} \sin\left(x^2\right)\,dx$$

and find error estimates for each value
that you calculate.

4.24   Use the Gaussian quadrature developed in Example 4.1 to find an approximate value for

$$\int_{0}^{3} \cos(8\sin(x-1)) + 1\,dx$$

How does your value compare to the other
estimates?

4.25   Consider

$$\int \frac{x^7 - 8x^4 + 2x + 5}{\sqrt{1 - x^2}}\,dx$$

   a) Use a computer algebra system
   to find an antiderivative.
   b) Use the fundamental theorem of
   calculus to evaluate the integral
   from -1 to 1.
   c) Compare your work to Example 4.2.

4.26   Use Gauss-Chebyshev quadrature
to evaluate

$$\int \frac{8x^4 + 5}{\sqrt{1 - x^2}}\,dx$$

Compare to Example 4.2 and explain your
result.

Fourier Series

4.27   Calculate the integrals
   a) $\displaystyle\int_{-\pi}^{\pi} \cos^2(nx)\,dx$
   b) $\displaystyle\int_{-\pi}^{\pi} \sin^2(nx)\,dx$

4.28   Find the Fourier series for the unit
step function.
   a) What value does the series converge to at $x = 0$?
   b) Plot several partial sums.

4.29   Prove that:
   a) The Fourier series of an even
   function has only cosine terms.

b) The Fourier series of an odd function has only sine terms.

4.30 Show that a function having *half-wave symmetry*, that is,

$$f(t) = -f(t + \pi)$$

has only odd harmonics; i.e., $a_{2n} = b_{2n} = 0$ for all $n \in \mathbb{N}$.

4.31 Verify the Fourier series listed in Table 4.6. Plot several partial sums for each series.

4.32 Compute the Fourier series of the *alternating pulse function*,

$$P(x) = \begin{cases} 0 & \dfrac{\pi}{2} \le |x| \le \pi \\ 1 & -\dfrac{\pi}{2} < x < 0 \\ -1 & 0 < x < \dfrac{\pi}{2} \end{cases}$$

Plot several partial sums.

In Exercises 4.33 to 4.36, assume that the function f has a continuous derivative bounded by M; i.e., $|f'(x)| \le M$.

4.33 Integrate the definition of a_k by parts to show that

$$a_k = -\frac{1}{k\pi} \int_{-\pi}^{\pi} f'(x) \sin(kx)\, dx$$

4.34 Show that

$$\left| \int_{-\pi}^{\pi} f'(x) \sin(kx)\, dx \right| \le 2\pi M$$

4.35 Combine the previous two exercises to show

$$|a_k| \le \frac{2M}{k}$$

4.36 Repeat the process of Exercises 4.33 to 4.35 for b_k to show

$$|b_k| \le \frac{2M}{k}$$

Conclude that the Fourier coefficients of a function having a bounded derivative go to zero as k goes to infinity.

4.37 Report on the *Gibbs phenomenon*. The phenomenon is sometimes called "ringing."

4.38 Investigate *Bessel's inequality*

$$\frac{a_0^2}{2} + \sum_{k=1}^{\infty} (a_k^2 + b_k^2) \le \frac{1}{\pi} \int_{-\pi}^{\pi} f^2(t)\, dt$$

4.39 Give a class presentation on the problem Fourier used to introduce Fourier series in his book *The Analytical Theory of Heat*.

Special Functions—The Gamma Function

4.40 What is $\lim\limits_{n \to \infty} \Gamma(-n + 1/2)$?

4.41 Calculate the value of
  a) $\Gamma(1/2)$
  b) $\Gamma(3/2)$
  c) $\Gamma(-1/2)$

4.42 Use a computer algebra system to produce a three-dimensional graph of $z = |\Gamma(x + yi)|$ over the domain $[-5, 5] \times [-5, 5]$. Explain the behavior of the plot.

4.43 Show that

$$\frac{\sqrt{2}}{2} \Gamma\left(\frac{1}{4}\right) \Gamma\left(\frac{3}{4}\right) = \pi$$

4.44 Consider $\Gamma(1/2)$.
  a) Write the integral form of $\Gamma(1/2)$.
  b) Change variables in the integral with $t = u^2$.

c) Determine the value of

$$\int_{-\infty}^{\infty} e^{-t^2} \, dt$$

d) How does this integral relate to probability?

4.45   Let $x = 15$ and $\epsilon = 10^{-4}$. How large an n guarantees that

$$\frac{x^n}{n!} < \epsilon$$

according to Stirling's formula?

4.46   Find the value of

$$\Gamma\left(\frac{2n+1}{2}\right)$$

for $n \in \mathbb{N}$.

4.47   Let $c > 0$ and $n \in \mathbb{N}$. Use Stirling's formula to prove

$$\lim_{n \to \infty} \frac{\Gamma(n+c)}{n^c \Gamma(n)} = 1$$

4.48   For $x > 0$, show that

$$\Gamma(x) = \int_0^1 \left[\ln\left(\frac{1}{t}\right)\right]^{x-1} dt$$

4.49   Verify the identities

a) $\dfrac{\sqrt{3}}{2} \displaystyle\prod_{k=1}^{3} \Gamma\left(\frac{k}{3}\right) = \pi$

b) $\dfrac{27}{8} \displaystyle\prod_{k=1}^{6} \Gamma\left(\frac{k}{3}\right) = \pi^2$

4.50   Show:

a) $\Gamma'(x) = \displaystyle\int_0^\infty t^{x-1} e^{-t} \ln(t) \, dt$

b) $\Gamma''(x) = \displaystyle\int_0^\infty t^{x-1} e^{-t} \ln^2(t) \, dt$

4.51   Find an expression for the *digamma function* $\Psi(x)$, the logarithmic derivative of $\Gamma(x)$,

$$\Psi(x) = \frac{\Gamma'(x)}{\Gamma(x)}$$

by differentiating under the integral sign. When is this valid?

4.52   Verify that

$$\Psi(1) - \Psi(1/2) = \ln(4)$$

Calculus Without Limits: Differential Algebra

4.53   Prove parts 2 and 3 of Theorem 4.14.
   [Hint: $D(\alpha/\beta) = D(\alpha \cdot (\beta)^{-1})$.]

4.54   Describe a procedure to decompose a rational function into a polynomial part and a proper fraction.

4.55   Expand $y = 1/(x^3 + 1)$ in partial fractions

a)   using only real numbers.

b)   using complex numbers.

4.56   Choose a very large number such as $n = 10^{1000}$. Let c be the nth digit in the decimal expansion of π. Does the integral $\int c e^{x^2}$ have an elementary antiderivative?

4.57   Report on Liouville's paper "Integration in finite terms."

4.58   Give a class presentation on *field extensions*.

The Closing Quote

4.59   Who said,

Life is good for only two things, discovering mathematics and teaching mathematics.

APPENDIX A

DEFINITIONS AND THEOREMS OF ELEMENTARY REAL ANALYSIS

The main definitions and results of elementary real analysis are collected here for reference. The theorems that underlie first-year calculus appear along with a number of interesting propositions that provide more depth and insight.

A.1   LIMITS

Definition A.1 (Accumulation Point) *Let $D \subseteq \mathbb{R}$. A point $a \in \mathbb{R}$ is an* accumulation point *of D iff every open interval containing a also contains a point $x \in D$ with $x \neq a$.*

Definition A.2 *Let $f : D \to \mathbb{R}$ and a be an accumulation point of D. Then*

$$\lim_{x \to a} f(x) = L$$

iff for every $\epsilon > 0$ there is a $\delta > 0$ such that whenever $x \in D$ and $0 < |x - a| < \delta$ then $|f(x) - L| < \epsilon$.

Introduction to Real Analysis. By William C. Bauldry
Copyright © 2009 John Wiley & Sons, Inc.

Theorem A.1 (Algebra of Limits) *Suppose that $f, g : D \to \mathbb{R}$ both have finite limits at $x = a$ an accumulation point of D and let $c \in \mathbb{R}$. Then*

- $\lim_{x \to a} c\, f(x) = c \lim_{x \to a} f(x)$

- $\lim_{x \to a} f(x) \pm g(x) = \lim_{x \to a} f(x) \pm \lim_{x \to a} g(x)$

- $\lim_{x \to a} f(x) \cdot g(x) = \lim_{x \to a} f(x) \cdot \lim_{x \to a} g(x)$

- *if $\lim_{x \to a} g(x) \neq 0$, then* $\lim_{x \to a} \dfrac{f(x)}{g(x)} = \dfrac{\lim_{x \to a} f(x)}{\lim_{x \to a} g(x)}$

Theorem A.2 ("Sandwich Theorem") *Suppose that $g(x) \leq f(x) \leq h(x)$ for all $x \in (a - h, a + h)$ for some $h > 0$. If $\lim_{x \to a} g(x) = L = \lim_{x \to a} h(x)$, then $\lim_{x \to a} f(x) = L$.*

A.2   CONTINUITY

Definition A.3 *Let $f : D \to \mathbb{R}$ and $a \in D$. Then f is continuous at $x = a$ iff a is an isolated point of D or for every $\epsilon > 0$ there is a $\delta > 0$ such that whenever $x \in D$ and $|x - a| < \delta$ then $|f(x) - f(a)| < \epsilon$.*

Theorem A.3 (Algebra of Continuity) *Suppose that $f, g : D \to \mathbb{R}$ both are continuous at $x = a \in D$ and that $c \in \mathbb{R}$. Then*

- *cf is continuous at a*

- *$f \pm g$ is continuous at a*

- *$f \cdot g$ is continuous at a*

- *if $g(a) \neq 0$, then f/g is continuous at a*

Corollary A.4 *Every real polynomial is continuous at every $x \in \mathbb{R}$.*

Theorem A.5 *If a function f is continuous at a and ϕ is a function such that $\lim_{t \to t_0} \phi(t) = a$, then*

$$\lim_{t \to t_0} f\left(\phi(t)\right) = f\left(\lim_{t \to t_0} \phi(t)\right)$$

Theorem A.6 (Continuity of Composition) *Let $f : A \to \mathbb{R}$ and $g : B \to \mathbb{R}$ where $f(A) \subseteq B$. Suppose that f is continuous at $x = a \in A$, that g is continuous at $x = f(a) \in B$. Then $g \circ f$ is continuous at $x = a$.*

Theorem A.7 *If a function f is continuous on a closed, bounded interval $[a, b]$, then f is bounded on $[a, b]$.*

Theorem A.8 (Intermediate Value Theorem) *If $f : [a, b] \to \mathbb{R}$ is continuous and if k is between $f(a)$ and $f(b)$, then there exists $c \in (a, b)$ such that $f(c) = k$.*

Corollary A.9 *Every odd-degree real polynomial has a real root.*

Corollary A.10 *Every real polynomial is a product of linear factors and irreducible (over $\mathbb{R}$) quadratic factors.*

Corollary A.11 (Fundamental Theorem of Algebra) *Every nth-degree real polynomial has n complex roots counting multiplicity.*

Theorem A.12 (Extreme Value Theorem) *If $f : [a, b] \to \mathbb{R}$ is continuous, then*

1. there exists $x_m \in [a, b]$ such that $f(x_m) = \min\limits_{x \in [a,b]} f(x)$

2. there exists $x_M \in [a, b]$ such that $f(x_M) = \max\limits_{x \in [a,b]} f(x)$

Definition A.4 (Uniform Continuity) *A function $f : D \to \mathbb{R}$ is* uniformly continuous *on $E \subseteq D$ iff for every $\epsilon > 0$ there is a $\delta > 0$ such that whenever $x_1, x_2 \in E$ and $|x_1 - x_2| < \delta$, then $|f(x_1) - f(x_2)| < \epsilon$.*

Theorem A.13 *If f is continuous on $[a, b]$, then f is uniformly continuous on $[a, b]$.*

Definition A.5 (Discontinuities) *If a function $f : D \to \mathbb{R}$ is not continuous at a point $a \in D$ and both $f(a+) = \lim_{x \to a+} f(x)$ and $f(a-) = \lim_{x \to a-} f(x)$ exist, then $x = a$ is a* simple discontinuity *of f. Further, if $f(a+) = f(a-) \neq f(a)$, then $x = a$ is a* removable discontinuity; *if $f(a+) \neq f(a-)$, then $x = a$ is a* jump discontinuity.*

A point of discontinuity that is not simple [either $f(a+)$ or $f(a-)$ fails to exist] is an essential discontinuity.*

Theorem A.14 *If $f : [a, b] \to \mathbb{R}$ is monotonic, then the set of points of discontinuity is countable.*

Theorem A.15 *The set of discontinuities of any function must be a countable union of closed sets or an F_σ set.*

Corollary A.16 *A function cannot be discontinuous only on the irrational reals.*

A.3 THE DERIVATIVE

Definition A.6 *Let $f : D \to \mathbb{R}$ and $a \in D$ be an accumulation point. Then*

$$f'(a) = \lim_{x \to a} \frac{f(x) - f(a)}{x - a}$$
$$= \lim_{h \to 0} \frac{f(a + h) - f(a)}{h}$$

Theorem A.17 *If f is differentiable at $x = a$, then f is continuous at $x = a$.*

Theorem A.18 (Algebra of Derivatives) *If $f, g : D \to \mathbb{R}$ are differentiable at $x = a$ and $c \in \mathbb{R}$, then at $x = a$,*

- $(cf)' = c(f')$

- $(f \pm g)' = f' \pm g'$

- $(f \cdot g)' = f' \cdot g + f \cdot g'$

- *if $g(a) \neq 0$, then* $\left(\dfrac{f}{g} \right)' = \dfrac{f' \cdot g + f \cdot g'}{g^2}$

Theorem A.19 (The Chain Rule) *Let $f : A \to B$ and $g : B \to \mathbb{R}$. Suppose that f is differentiable at $x = a \in A$ and that g is differentiable at $x = b = f(a) \in B$. Then $g \circ f$ is differentiable at $x = a$ and*

$$(g \circ f)'(a) = g'(f(a)) \cdot f'(a)$$

Corollary A.20 *Let u be a differentiable function of x and $r \in \mathbb{R}$. Then, when defined,*

$$(u^r)' = r\, u^{r-1} \cdot u'$$

$$(e^u)' = e^u \cdot u' \qquad\qquad \ln(u)' = \frac{1}{u} \cdot u'$$

$$\sin(u)' = \cos(u) \cdot u' \qquad\qquad \cos(u)' = -\sin(u) \cdot u'$$

$$\tan(u)' = \sec^2(u) \cdot u' \qquad\qquad \cot(u)' = -\csc^2(u) \cdot u'$$

$$\sec(u)' = \sec(u)\tan(u) \cdot u' \qquad \csc(u)' = -\csc(u)\cot(u) \cdot u'$$

$$\sin^{-1}(u)' = \frac{1}{\sqrt{1-u^2}} \cdot u' \qquad \cos^{-1}(u)' = \frac{-1}{\sqrt{1-u^2}} \cdot u'$$

$$\tan^{-1}(u)' = \frac{1}{1+u^2} \cdot u' \qquad \cot^{-1}(u)' = \frac{-1}{1+u^2} \cdot u'$$

$$\sec^{-1}(u)' = \frac{1}{|u|\sqrt{u^2-1}} \cdot u' \qquad \csc^{-1}(u)' = \frac{-1}{|u|\sqrt{u^2-1}} \cdot u'$$

Theorem A.21 (Inverse Function Theorem) *Let $f : [a, b] \to \mathbb{R}$ be differentiable with $f'(x) \neq 0$ for any $x \in [a, b]$. Then*

- *f is injective (1–1)*

- *f^{-1} is continuous on $f([a, b])$*

- *f^{-1} is differentiable on $f([a, b])$*

- *$(f^{-1})'(y) = \dfrac{1}{f'(x)}$ where $y = f(x)$*

Theorem A.22 *Suppose that $f : [a, b] \to \mathbb{R}$ has an extremum at $c \in (a, b)$. If f is differentiable at $c \in (a, b)$, then $f'(c) = 0$.*

Theorem A.23 (Rolle's Theorem) *If $f : D \to \mathbb{R}$ is continuous on $[a, b] \subseteq D$ and differentiable on (a, b) with $f(a) = f(b)$, then there exists a value $c \in (a, b)$ such that $f'(c) = 0$.*

Theorem A.24 (Mean Value Theorem) *If $f : D \to \mathbb{R}$ is continuous on $[a, b]$ and differentiable on (a, b), then there exists a value $c \in (a, b)$ such that*

$$\frac{f(b) - f(a)}{b - a} = f'(c)$$

Corollary A.25 *If $f : D \to \mathbb{R}$ is continuous on $[a, b]$ and differentiable on (a, b), then there exists a value $\theta \in (0, 1)$ such that*

$$f(a + h) = f(a) + h \cdot f'(a + \theta h)$$

Corollary A.26 (Racetrack Principle) *If $f : D \to \mathbb{R}$ is continuous on $[a, b]$, differentiable on (a, b), and $m \leq f'(x) \leq M$, then*

$$f(a) + m \cdot (b - a) \leq f(b) \leq f(a) + M \cdot (b - a)$$

Corollary A.27 *If $f : D \to \mathbb{R}$ is continuous on $[a, b]$, differentiable on (a, b), and $f'(x) = 0$ on D, then f is a constant function.*

Corollary A.28 *If $f, g : D \to \mathbb{R}$ are continuous on $[a, b]$, differentiable on (a, b), and $f'(x) = g'(x)$ on D, then $f(x) = g(x) + k$ on D where k is a constant.*

Corollary A.29 *If $f : D \to \mathbb{R}$ is differentiable on $[a, b]$, then f' has the* intermediate value property.

Theorem A.30 (Cauchy's Mean Value Theorem) *If $f, g : D \to \mathbb{R}$ are continuous on $[a, b]$ and differentiable on (a, b), then there exists a value $c \in (a, b)$ such that*

$$f'(c) [g(b) - g(a)] = g'(c) [f(b) - f(a)]$$

or, when denominators are nonzero,

$$\frac{f(b) - f(a)}{g(b) - g(a)} = \frac{f'(c)}{g'(c)}$$

Theorem A.31 (Darboux Intermediate Value Theorem) *Let $f : D \to \mathbb{R}$ be continuous on $[a, b]$ and differentiable on (a, b), from the right at a and the left at b. Then f' has the intermediate value property; i.e., for every t between $f'_+(a)$ and $f'_-(b)$, there is an $\hat{x} \in [a, b]$ with $f(\hat{x}) = t$.*

Definition A.7 (Uniform Differentiability) *Let $f : [a, b] \to \mathbb{R}$. Then f is* uniformly differentiable *on $[a, b]$ iff f is differentiable on $[a, b]$ and, for every $\epsilon > 0$, there exists a $\delta > 0$ such that whenever $x_1, x_2 \in [a, b]$ with $|x_1 - x_2| < \delta$ it must follow that*

$$\left| \frac{f(x_1) - f(x_2)}{x_1 - x_2} - f'(x_1) \right| < \epsilon$$

Corollary A.32 *If $f : D \to \mathbb{R}$ is uniformly differentiable on $[a, b]$, then f' is continuous on $[a, b]$.*

Definition A.8 (Lipschitz Condition) *Let $f : D \to \mathbb{R}$. If there are positive constants M and α such that for any $x_1, x_2 \in D$*

$$|f(x_1) - f(x_2)| \leq M \cdot |x_1 - x_2|^{\alpha}$$

then f is Lipschitz-α *with constant M, written $f \in \mathrm{Lip}_M \, \alpha$.*

Theorem A.33 *If $f \in \mathrm{Lip}_M \, \alpha$ on D, then*

1. f is continuous,

2. if $\alpha > 1$, f is constant.

Corollary A.34 *If $f : [a, b] \to \mathbb{R}$ is differentiable, then $f \in \mathrm{Lip}_M \, 1$.*

Definition A.9 (Measure Zero) *A set E has* measure zero *if and only if for any $\epsilon > 0$ the set E can be covered by a countable collection of open intervals having total length less than ϵ; i.e., $E \subseteq \bigcup_i (a_i, b_i)$ where $\sum_i (b_i - a_i) < \epsilon$.*

Definition A.10 (Almost Everywhere) *A property P holds* almost everywhere *if the set $\{x : P(x) \text{ is not true}\}$ has measure zero.*

Theorem A.35 (Rademacher's Theorem) *If $f \in \mathrm{Lip}_M \, 1$, then f is differentiable almost everywhere.*

Theorem A.36 (Lebesgue Differentiation Theorem) *If $f : \mathbb{R} \to \mathbb{R}$ is continuous and nondecreasing, then f is differentiable almost everywhere.*

Definition A.11 (Higher Order Derivatives) *The nth derivative of $f(x)$, if it exists, is given by*

$$f^{(n)}(x) = \frac{d}{dx} f^{(n-1)}(x)$$

for $n > 1$ where $f^{(0)} = f$.

Theorem A.37 *Let $f : D \to \mathbb{R}$ be m times continuously differentiable. Then f has a root of multiplicity m at $x = r$ iff $f^{(m)}(r) \neq 0$, but*

$$f(r) = f'(r) = \cdots = f^{(m-1)}(r) = 0$$

Theorem A.38 (First Derivative Test for Extrema) *Let f be continuous on $[a, b]$ and differentiable on (a, b). If $f'(c) = 0$ for $c \in (a, b)$ and f' changes sign from*

- *negative to positive around c, then c is a relative minimum of f;*

- *positive to negative around c, then c is a relative maximum of f.*

If f' does not change sign around c, then c is a stationary "terrace point" of f.

Theorem A.39 (Second Derivative Test for Extrema) *Let f be continuous on $[a, b]$ and twice differentiable on (a, b). If $f'(c) = 0$ for $c \in (a, b)$ and*

- $f''(c)$ *is positive, then c is a relative minimum of f;*

- $f''(c)$ *is negative, then c is a relative maximum of f;*

- $f''(c) = 0$, *then the test fails.*

Theorem A.40 (Taylor's Theorem or Extended Law of the Mean) *Let $n \in \mathbb{N}$ and suppose that f has $n + 1$ derivatives on $(a - h, a + h)$ for some $h > 0$. Then for $x \in (a - h, a + h)$*

$$f(x) = f(a) + \sum_{k=1}^{n} \frac{f^{(k)}(a)}{k!} (x - a)^k + R_n(x)$$

where

$$R_n(x) = \frac{f^{(n+1)}(c)}{(n + 1)!} (x - a)^{n+1}$$

for some c between x and a.

Theorem A.41 (Taylor's Theorem with Lagrange's Form of the Remainder) *Let $n \in \mathbb{N}$ and suppose that f has $n + 1$ continuous derivatives on $(a - h, a + h)$ for some $h > 0$. Then for $x \in (a - h, a + h)$*

$$f(x) = f(a) + \sum_{k=1}^{n} \frac{f^{(k)}(a)}{k!} (x - a)^k + L_n(x)$$

where

$$L_n(x) = \frac{1}{n!} \int_a^x f^{(n+1)}(t) \cdot (x - t)^n \, dt$$

Theorem A.42 (Bernstein) *Let I be an interval and $f : I \to \mathbb{R}$ have derivatives of all orders. If f and all its derivatives are nonnegative, then the Taylor series of f converges on I.*

Theorem A.43 (L'Hôpital's Rule) *Suppose that f and g are differentiable on an open interval I containing a and that*

$$\lim_{x \to a} f(x) = 0 = \lim_{x \to a} g(x)$$

while $g'(x) \neq 0$ on I. Then, if the limit exists,

$$\lim_{x \to a} \frac{f'(x)}{g'(x)} = \lim_{x \to a} \frac{f(x)}{g(x)}$$

Corollary A.44 *Let $n \in \mathbb{N}$. Then*

$$\lim_{x \to \infty} \frac{x^n}{e^x} = 0 \quad and \quad \lim_{x \to \infty} \frac{\ln(x)}{\sqrt[n]{x}} = 0$$

Corollary A.45 *If f is twice differentiable on an open interval I and $x \in I$, then*

$$f''(x) = \lim_{h \to 0} \frac{f(x+h) - 2f(x) + f(x-h)}{h^2}$$

A.4   RIEMANN INTEGRATION

Definition A.12 (Partition) *A partition P of a closed interval $[a, b]$ is an ordered set of values $\{x_i \mid i = 0, \ldots, n\}$ such that $a = x_0 < x_1 < \cdots < x_n = b$. The norm or mesh of the partition is*

$$\|P\| = \max\{\Delta x_i \mid i = 1 \ldots n\}$$

where $\Delta x_i = x_i - x_{i-1}$.

Definition A.13 (Cauchy Sum) *Let $f : [a, b] \to \mathbb{R}$ be a continuous function and P be a partition of $[a, b]$. Then the Cauchy sum (or "left endpoint sum") of f (w.r.t. P) is*

$$C(P, f) = \sum_{k=1}^{n} f(x_{k-1}) \Delta x_k$$

Definition A.14 (Riemann Sum) *Let $f : [a, b] \to \mathbb{R}$ be a bounded function, P be a partition of $[a, b]$, and $c_k \in [x_{k-1}, x_k]$ for each k. Then the Riemann sum of f (w.r.t. P and $\{c_k\}$) is*

$$R(P, f) = \sum_{k=1}^{n} f(c_k) \Delta x_k$$

Definition A.15 (Darboux Sums) *Let $f : [a, b] \to \mathbb{R}$ and P be a partition of $[a, b]$ and set*

$$M_k(f) = \sup_{[x_{k-1}, x_k]} f(x) \quad and \quad m_k(f) = \inf_{[x_{k-1}, x_k]} f(x)$$

Then the upper and lower Darboux sums of f (w.r.t. P) are

$$U(P, f) = \sum_{k=1}^{n} M_k \Delta x_k \quad and \quad L(P, f) = \sum_{k=1}^{n} m_k \Delta x_k$$

respectively.

Lemma A.46 *Let* $f : [a, b] \to \mathbb{R}$ *be a bounded function, say, below by* m *and above by* M, *and let* P *be a partition of* $[a, b]$. *Then*

$$m(b - a) \leq L(P, f) \leq R(P, f) \leq U(P, f) \leq M(b - a)$$

for all choices of $\{c_k\}$.

Lemma A.47 *Suppose* $f : [a, b] \to \mathbb{R}$ *is a bounded function and* P *and* Q *are partitions of* $[a, b]$. *If* $P \subseteq Q$ (*i.e.,* Q *is a* finer *partition*), *then*

1. $L(P, f) \leq L(Q, f)$ *and* $U(Q, f) \leq U(P, f)$

2. $L(P, f) \leq U(Q, f)$

Definition A.16 *Set*

$$\overline{\int_a^b} f(x) \, dx = \inf_P U(P, f)$$

and

$$\underline{\int_a^b} f(x) \, dx = \sup_P L(P, f)$$

Lemma A.48 *Let* $f : [a, b] \to \mathbb{R}$ *be a bounded function and* P *be a partition of* $[a, b]$. *Then*

$$L(P, f) \leq \underline{\int_a^b} f(x) \, dx \leq \overline{\int_a^b} f(x) \, dx \leq U(P, f)$$

Definition A.17 *A bounded function* $f : [a, b] \to \mathbb{R}$ *is* Riemann integrable *on* $[a, b]$ *if and only if*

$$\underline{\int_a^b} f(x) \, dx = \overline{\int_a^b} f(x) \, dx = \int_a^b f(x) \, dx = A \in \mathbb{R}$$

Set $\mathfrak{R}[a, b] = \{f \mid f$ *is Riemann integrable on* $[a, b]\}$.

Theorem A.49 *The bounded function* $f : [a, b] \to \mathbb{R}$ *is Riemann integrable on* $[a, b]$ *if and only if for any* $\epsilon > 0$ *there is a partition* P *such that* $U(P, f) - L(P, f) < \epsilon$.

Theorem A.50 *If* f *is monotone on* $[a, b]$, *then* $f \in \mathfrak{R}[a, b]$.

Theorem A.51 *If* f *is continuous on* $[a, b]$, *then* $f \in \mathfrak{R}[a, b]$.

Theorem A.52 *If* $f, g \in \mathfrak{R}[a, b]$ *and* $c \in \mathbb{R}$, *then*

1. $f + g \in \mathfrak{R}[a, b]$ *and* $\int_a^b (f + g)(x) \, dx = \int_a^b f(x) \, dx + \int_a^b g(x) \, dx$

2. $cf \in \mathfrak{R}[a, b]$ *and* $\int_a^b cf(x) \, dx = c \int_a^b f(x) \, dx$

Theorem A.53 *If $f, g \in \mathfrak{R}[a, b]$ and $f(x) \le g(x)$ on $[a, b]$, then*

$$\int_a^b f(x)\, dx \le \int_a^b g(x)\, dx$$

Theorem A.54 *If $f \in \mathfrak{R}[a, b]$, then $|f| \in \mathfrak{R}[a, b]$ and*

$$\left| \int_a^b f(x)\, dx \right| \le \int_a^b |f(x)|\, dx$$

Theorem A.55 *Let $f : [a, b] \to \mathbb{R}$ is bounded and $c \in (a, b)$. Then $f \in \mathfrak{R}[a, b]$ if and only if $f \in \mathfrak{R}[a, c]$ and $f \in \mathfrak{R}[c, b]$ and further*

$$\int_a^b f(x)\, dx = \int_a^c f(x)\, dx + \int_c^b f(x)\, dx$$

Theorem A.56 *If $f \in \mathfrak{R}[a, b]$ and g is continuous on $f([a, b])$, then $g \circ f \in \mathfrak{R}[a, b]$.*

Corollary A.57 *Let $f, g \in \mathfrak{R}[a, b]$ and $n \in \mathbb{N}$. Then*

1. $f^n \in \mathfrak{R}[a, b]$

2. $f \cdot g \in \mathfrak{R}[a, b]$

Lemma A.58 *Let $f \in \mathfrak{R}[a, b]$ and $c, d \in [a, b]$. Then*

1. $\displaystyle \int_c^c f(x)\, dx = 0$

2. $\displaystyle \int_c^d f(x)\, dx = -\int_d^c f(x)\, dx$

Theorem A.59 (Bonnet's Theorem or First Mean Value Theorem for Integrals)
Let f be continuous on $[a, b]$ and $0 \le g \in \mathfrak{R}[a, b]$. Then there exists a value $c \in (a, b)$ such that

$$\int_a^b f(x)\, g(x)\, dx = f(c) \int_a^b g(x)\, dx$$

Theorem A.60 (Second Mean Value Theorem for Integrals) *If f is a monotone function on $[a, b]$, then there exists a value $c \in (a, b)$ such that*

$$\int_a^b f(x)\, dx = f(a)(c - a) + f(b)(b - c)$$

Theorem A.61 (The Fundamental Theorem of Calculus)

- *Define $F(x) = \displaystyle \int_a^x f(t)\, dt$. Then F is continuous and, if f is continuous at x_0, then $F'(x_0) = f(x_0)$.*

- *If $F' = f$ on $[a, b]$, then $\displaystyle \int_a^b f(x)\, dx = F(b) - F(a)$.*

Theorem A.62 (Midpoint Rule for Numerical Integration) *Suppose* $f : [a, b] \to \mathbb{R}$ *is integrable and that* $|f''(x)| \le M_2$ *on* $[a, b]$. *For* n *given, set* $x_k = a + k(b-a)/n$ *with* $k = 0, \ldots, n$ *and set* $\bar{y}_k = f\left((x_k + x_{k+1})/2\right)$ *for* $k = 0, \ldots, n - 1$. *Then*

$$\int_a^b f(x)\, dx \approx \frac{b-a}{n} \left(\bar{y}_0 + \bar{y}_1 + \cdots + \bar{y}_{n-1}\right)$$

with the absolute value of the error bounded by

$$|error| \le \frac{1}{24} \frac{(b-a)^3}{n^2} M_2$$

Theorem A.63 (Trapezoid Rule) *Suppose* $f : [a, b] \to \mathbb{R}$ *is integrable and that* $|f''(x)| \le M_2$ *on* $[a, b]$. *For* n *given, set* $x_k = a + k(b-a)/n$ *and set* $y_k = f(x_k)$ *with* $k = 0, \ldots, n$. *Then*

$$\int_a^b f(x)\, dx \approx \frac{b-a}{2n} \left(y_0 + 2y_1 + 2y_2 + \cdots + 2y_{n-1} + y_n\right)$$

with the absolute value of the error bounded by

$$|error| \le \frac{1}{12} \frac{(b-a)^3}{n^2} M_2$$

Theorem A.64 (Simpson's Rule) *Suppose* $f : [a, b] \to \mathbb{R}$ *is integrable and that* $\left|f^{(4)}(x)\right| \le M_4$ *on* $[a, b]$. *For* n *even, set* $x_k = a + k(b-a)/n$ *and set* $y_k = f(x_k)$ *with* $k = 0, \ldots, n$. *Then*

$$\int_a^b f(x)\, dx \approx \frac{b-a}{3n} \left(y_0 + 4y_1 + 2y_2 + 4y_3 + 2y_4 + \cdots + 4y_{n-1} + y_n\right)$$

with the absolute value of the error bounded by

$$|error| \le \frac{1}{180} \frac{(b-a)^5}{n^4} M_4$$

Theorem A.65 (Cauchy-Bunyakovsky-Schwarz Inequality) *If* $f, g \in \mathfrak{R}[a, b]$, *then*

$$\left[\int_a^b f(x)\, g(x)\, dx\right]^2 \le \left[\int_a^b f^2(x)\, dx\right] \cdot \left[\int_a^b g^2(x)\, dx\right]$$

A.5 RIEMANN-STIELTJES INTEGRATION

Definition A.18 *Let* α *be a monotonically increasing function on* $[a, b]$. *For any partition* P *define* $\Delta\alpha_i = \alpha(x_i) - \alpha(x_{i-1})$.

Definition A.19 (Upper and Lower Riemann-Stieltjes Integrals) *Let f be a function that is bounded and α be monotonically increasing on $[a, b]$. For each partition P, define the* upper *and* lower *Riemann-Stieltjes sums by*

$$U(P, f, \alpha) = \sum_{k=1}^{n} M_k \Delta \alpha_k$$

$$L(P, f, \alpha) = \sum_{k=1}^{n} m_k \Delta \alpha_k$$

Now, define the upper *and* lower *Riemann-Stieltjes integrals as*

$$\overline{\int_a^b} f(x)\, d\alpha(x) = \inf_P U(P, f, \alpha)$$

$$\underline{\int_a^b} f(x)\, d\alpha(x) = \inf_P L(P, f, \alpha)$$

Definition A.20 *If $\overline{\int} f d\alpha = \underline{\int} f d\alpha$, then f is* Riemann-Stieltjes integrable *and we write $f \in \mathfrak{R}(\alpha)$.*

Theorem A.66 *A function f is Riemann-Stieltjes integrable on $[a, b]$ if and only if for every $\epsilon > 0$ there is a partition P of $[a, b]$ such that*
$$U(P, f, \alpha) - L(P, f, \alpha) < \epsilon$$

Theorem A.67 *If f is continuous on $[a, b]$, then $f \in \mathfrak{R}(\alpha)$ on $[a, b]$.*

Theorem A.68 *If f is monotonic on $[a, b]$ and α is continuous, then $f \in \mathfrak{R}(\alpha)$ on $[a, b]$.*

Theorem A.69 *If f is bounded and has finitely many discontinuities on $[a, b]$ and α is continuous at each discontinuity of f, then $f \in \mathfrak{R}(\alpha)$.*

Theorem A.70 *If f is bounded on $[a, b]$ and $f \in \mathfrak{R}(\alpha)$, then there exists m and $M \in \mathbb{R}$ such that*

$$m(\alpha(b) - \alpha(a)) \leq \int_a^b f(x)\, d\alpha(x) \leq M(\alpha(b) - \alpha(a))$$

Theorem A.71 *Let f and $g \in \mathfrak{R}(\alpha)$ on $[a, b]$ and $c \in \mathbb{R}$. Then*

1. $\int_a^b cf(x)\, d\alpha(x) = c \int_a^b f(x)\, d\alpha(x)$
2. $\int_a^b f(x)\, d\, c\alpha(x) = c \int_a^b f(x)\, d\alpha(x)$
3. $\int_a^b (f + g)(x)\, d\alpha(x) = \int_a^b f(x)\, d\alpha(x) + \int_a^b g(x)\, d\alpha(x)$
4. $\int_a^b f(x)\, d(\alpha_1(x) + \alpha_2(x)) = \int_a^b f(x)\, d\alpha_1(x) + \int_a^b f(x)\, d\alpha_2(x)$
5. $f \cdot g \in \mathfrak{R}(\alpha)$

Theorem A.72 *Let f and $g \in \mathfrak{R}(\alpha)$ on $[a, b]$. If $f(x) \leq g(x)$, then*

$$\int_a^b f(x)\, d\alpha(x) \leq \int_a^b g(x)\, d\alpha(x)$$

Theorem A.73 *If $f \in \mathfrak{R}(\alpha)$, then*

$$\left| \int_a^b f(x)\, d\alpha(x) \right| \leq \int_a^b |f(x)|\, d\alpha(x)$$

Definition A.21 *Define the* Heaviside function *to be*

$$U(x) = \begin{cases} 0 & x \leq 0 \\ 1 & x > 0 \end{cases}$$

Theorem A.74 *If f is bounded on $[a, b]$ and continuous at $x_0 \in (a, b)$, then*

$$\int_a^b f(x)\, dU(x - x_0) = f(x_0)$$

Theorem A.75 *Let $c_n \geq 0$ with $\sum_n c_n$ converging and $\{x_n\}$ be a sequence of distinct points in (a, b). Define*

$$\alpha(x) = \sum_{k=1}^{\infty} c_k U(x - x_k)$$

Let f be continuous on $[a, b]$. Then

$$\int_a^b f(x)\, d\alpha(x) = \sum_{k=1}^{\infty} c_k f(x_k)$$

Theorem A.76 *If f is bounded and $\alpha' \in \mathfrak{R}(\alpha)$, then $f \in \mathfrak{R}(\alpha)$ iff $f\alpha' \in \mathfrak{R}$ and*

$$\int_a^b f(x)\, d\alpha(x) = \int_a^b f(x)\alpha'(x)\, dx$$

Theorem A.77 (Hölder's Inequality) *Let f and g be in $\mathfrak{R}(\alpha)$ and let $p, q > 0$ be such that $1/p + 1/q = 1$. Then*

$$\left| \int_a^b f(x)g(x)\, d\alpha(x) \right| \leq \left[\int_a^b |f(x)|^p d\alpha(x) \right]^{1/p} \left[\int_a^b |g(x)|^q d\alpha(x) \right]^{1/q}$$

If $p = 2$, this is called the Cauchy-Bunyakovski-Schwarz inequality.

Theorem A.78 (Minkowski's Inequality) *Let $p > 1$ and let f^p and g^p be in $\mathfrak{R}(\alpha)$. Then*

$$\left[\int_a^b [f(x) + g(x)]^p\, d\alpha(x) \right]^{1/p} \leq \left[\int_a^b |f(x)|^p d\alpha(x) \right]^{1/p} + \left[\int_a^b |g(x)|^p d\alpha(x) \right]^{1/p}$$

A.6 SEQUENCES AND SERIES OF CONSTANTS

Definition A.22 *A real-valued sequence is a function* $a : \mathbb{N} \to \mathbb{R}$. *Sequence terms are denoted by* $a(n) = a_n$.

Definition A.23 *Let* $f : \mathbb{N} \to \mathbb{N}$ *be a strictly increasing function. The composition* $a \circ f$ *forms a* subsequence *and is denoted by* $a(f(k)) = a_{n_k}$.

Definition A.24 *A sequence* converges, *written* $\lim_{n\to\infty} a_n = L$, *iff for every* $\epsilon > 0$ *there is an* $N \in \mathbb{N}$ *such that, if* $n > N$, *then* $|a_n - L| < \epsilon$.

Theorem A.79 (Corollary to the Heine-Borel Theorem) *A bounded sequence has a convergent subsequence.*

Definition A.25 (Cauchy Sequence) *A sequence* a_n *is* Cauchy *iff, for every* $\epsilon > 0$, *there is an* $N \in \mathbb{N}$ *such that, if* $n, m > N$, *then* $|a_n - a_m| < \epsilon$.

Theorem A.80 *A sequence is Cauchy iff the sequence converges.*

Definition A.26 *A sequence* $\{a_n\}$ *is*

- monotonically increasing *iff* $a_n \leq a_{n+1}$ *for all* n,
- monotonically decreasing *iff* $a_n \geq a_{n+1}$ *for all* n.

Theorem A.81 *If* $\{a_n\}$ *is monotonic, then* $\{a_n\}$ *converges iff it is bounded.*

Definition A.27 *Let* $\{a_n\}$ *be a sequence. Define*

- $\displaystyle \limsup_{n\to\infty} a_n = \lim_{n\to\infty} \left(\sup_{k\geq n} a_k \right)$ *which is also equal to* $\displaystyle \inf_{n\geq 0} \left(\sup_{k\geq n} a_k \right)$

- $\displaystyle \liminf_{n\to\infty} a_n = \lim_{n\to\infty} \left(\inf_{k\geq n} a_k \right)$ *which is also equal to* $\displaystyle \sup_{n\geq 0} \left(\inf_{k\geq n} a_k \right)$

Theorem A.82 *Let* $\{a_n\}$ *be a real-valued sequence. Then* $\lim_{n\to\infty} a_n = a$ *iff* $\liminf_{n\to\infty} a_n = a = \limsup_{n\to\infty} a_n$.

Definition A.28 *Let* $\{a_n\}$ *be a sequence. The associated* series *is* $s_n = \sum_{k=1}^{n} a_n$. *The terms* s_n *are called the* partial sums *of the series. The series converges to* s, *written as* $s = \sum_{k=1}^{\infty} a_n$, *iff the sequence of partial sums* $\{s_n\}$ *converges to* s.

Theorem A.83 *If* $s_n = \sum_{k=1}^{\infty} a_n$ *converges, then* $\lim_{n\to\infty} a_n = 0$.

Definition A.29 (Cauchy Series) *A series* $s_n = \sum_{k=1}^{\infty} a_n$ *is called* Cauchy *iff for every* $\epsilon > 0$ *there is an* $N \in \mathbb{N}$ *such that, if* $n \geq m > N$, *then*

$$\left| \sum_{k=m+1}^{n} a_k \right| < \epsilon$$

Theorem A.84 *Every Cauchy series in $\mathbb{R}$ converges.*

Theorem A.85 (The Comparison Test) *Let $\sum_n a_n$ be a series.*

- *If $|a_n| \le c_n$ for all $n \ge N_0$ and $\sum c_n$ converges, then $\sum a_n$ converges.*

- *If $a_n \ge d_n > 0$ for all $n \ge N_0$ and $\sum d_n$ diverges, then $\sum a_n$ diverges.*

Theorem A.86 (The Limit Comparison Test) *Let $\sum_n a_n$ and $\sum_n b_n$ be positive series. If $0 < \lim_{n \to \infty} a_n/b_n < \infty$, then the series either both converge or both diverge.*

Theorem A.87 (Cauchy Condensation Test) *Let $\{a_n\}$ be a nonnegative decreasing sequence. Then the series $\sum_n a_n$ converges if and only if $\sum_n (2^n a_{2^n})$ converges.*

Theorem A.88 (The Ratio Test, I) *The series $\sum_n a_n$ converges if*

$$\limsup_n \left| \frac{a_{n+1}}{a_n} \right| < 1$$

and diverges if $|a_{n+1}/a_n| > 1$ for $n \ge N_0$.

Corollary A.89 (The Ratio Test, II) *For the series $\sum_n a_n$, define*

$$\rho = \lim_n \left| \frac{a_{n+1}}{a_n} \right|$$

Then,

- *If $\rho < 1$, the series converges.*

- *If $\rho > 1$, the series diverges.*

- *If $\rho = 1$, the test fails.*

Theorem A.90 (The Root Test) *For the series $\sum_n a_n$, define*

$$\rho = \limsup_n \sqrt[n]{|a_n|}$$

Then,

- *If $\rho < 1$, the series converges.*

- *If $\rho > 1$, the series diverges.*

- *If $\rho = 1$, the test fails.*

Theorem A.91 (The Integral Test) *Let $f : [1, \infty) \to \mathbb{R}$ be a continuous, positive, decreasing function such that $f(n) = a_n$. Then the improper integral $\int_1^\infty f(x)\,dx$ converges if and only if the series $\sum_n a_n$ converges.*

Theorem A.92 (Alternating Series Test) *If $a_n \ge a_{n+1} > 0$ for all $n \in \mathbb{N}$ and $\lim_{n \to \infty} a_n = 0$, then $\sum_n (-1)^n a_n$ converges.*

A.7 SEQUENCES AND SERIES OF FUNCTIONS

Definition A.30 (Pointwise Convergence) *A sequence of functions* $f_n : E \to \mathbb{R}$ *converges pointwise on* E *iff for each* $x \in E$ *the sequence* $\{f_n(x)\}$ *converges.*

Definition A.31 (Convergence in Mean) *A sequence of integrable real functions* $f_n : [a, b] \to \mathbb{R}$ *converges in mean to* f *iff*

$$\lim_{n \to \infty} \left[\int_a^b [f_n(x) - f(x)]^2 dx \right]^{1/2} = 0$$

Definition A.32 (Uniform Convergence) *A sequence of functions* $f_n : E \to \mathbb{R}$ *converges uniformly to* f *on* E *iff for every* $\epsilon > 0$ *there is an* $N \in \mathbb{N}$ *such that whenever* $n > N$ *then* $|f_n(x) - f(x)| < \epsilon$ *for all* $x \in E$.

Definition A.33 (Cauchy Criterion for Uniform Convergence) *A sequence of functions* $f_n : E \to \mathbb{R}$ *converges uniformly on* E *iff for every* $\epsilon > 0$ *there is an* $N \in \mathbb{N}$ *such that whenever* $n, m > N$ *then* $|f_n(x) - f_m(x)| < \epsilon$ *for all* $x \in E$.

Theorem A.93 *Let* $f_n : E \to \mathbb{R}$ *be a sequence of functions converging uniformly to* f *on* E. *If each* f_n *is continuous on* E, *then* f *is continuous on* E.

Theorem A.94 *Let* $f_n : E \to \mathbb{R}$ *be a sequence of functions converging uniformly to* f *on* E. *Then*

$$\lim_{n \to \infty} \lim_{x \to a} f_n(x) = \lim_{x \to a} \lim_{n \to \infty} f_n(x)$$

for $x, a \in E$.

Theorem A.95 *Let* $f_n : [a, b] \to \mathbb{R}$ *be a sequence of integrable functions converging uniformly on* $[a, b]$. *Then*

$$\lim_{n \to \infty} \int_a^b f_n(x)\, dx = \int_a^b \lim_{n \to \infty} f_n(x)\, dx$$

Theorem A.96 (Dini's Theorem) *Let* $f_n : [a, b] \to \mathbb{R}$ *be a monotonic sequence of functions converging pointwise to* f *on* $[a, b]$ *where* $-\infty < a < b < \infty$. *Then* f_n *converges uniformly to* f *on* $[a, b]$.

Theorem A.97 *Let* $f_n : [a, b] \to \mathbb{R}$ *be a sequence of continuously differentiable functions converging pointwise to* f *on* $[a, b]$. *If* f_n' *converges uniformly on* $[a, b]$, *then* f_n *converges uniformly to* f *and* f_n' *converges uniformly to the continuous function* f' *on* $[a, b]$.

Theorem A.98 (The Weierstrass M-test) *Let* $\sum_n f_n(x)$ *be a series of functions all defined on* $D \subseteq \mathbb{R}$. *If there is a convergent series of constants* $\sum_n M_n$ *such that for each* n *we have* $|f_n(x)| \le M_n$ *for all* $x \in D$, *then* $\sum_n f_n(x)$ *converges both uniformly and absolutely on* D.

Theorem A.99 (Abel's Uniform Convergence Test) *Let* $f_n : D \to \mathbb{R}$ *be a bounded, monotonically decreasing sequence of functions and* $\sum_n a_n$ *be a convergent series of constants. Then* $\sum_n a_n f_n(x)$ *converges uniformly on* D.

APPENDIX B

A BRIEF CALCULUS CHRONOLOGY

Few new branches of mathematics are the work of single individuals. . . . Far less is the development of the calculus to be ascribed to one or two men.

— Carl Boyer

Many elementary calculus students have the impression that the subject began and ended with Newton and Leibniz. A timeline of events in the development of calculus can illustrate Newton's statement "I have stood on the shoulders of giants" and show analysis is still a being developed today. Each name in the timeline can lead to a historical report on a step in the development.

Introduction to Real Analysis. By William C. Bauldry
Copyright © 2009 John Wiley & Sons, Inc.

Figure B.1 Sir Isaac Newton (1643–1727). Portrait by Sir Godfrey Kneller, 1689.

Figure B.2 Gottfried Wilhelm von Leibniz (1646–1716). *Courtesy WPClipArt.*

Hippocrates of Chios	c. 440 BC	Quadrature of the lune
Eudoxus of Cnidus	c. 370 BC	Method of exhaustion
Archimedes of Syracuse	c. 225 BC	Quadrature of the parabola
Al-Khwarizmi	c. 850	*Hisab al-jabr w'al-muqabala*—first algebra text
François Viète	1591	Systematic algebraic notation
Johannes Kepler	1615	Calculating volumes from infinitely many cones
Pierre de Fermat	1629	Determining maxima and minima of curves
Blaise Pascal	1650	Quadrature of polynomials
Isaac Barrow	1664	Differential triangle
Isaac Newton	1666	*Annus Mirabilis*, fluxions and fluents
James Gregory	1671	Taylor series
Gottfried Leibniz	1672	Integration
Michel Rolle	1691	Rolle's theorem
Johann Bernoulli	1694	l'Hôpital's Rule
Marquis de l'Hôpital	1696	First calculus textbook
Leonhard Euler	1750	Infinite sums, differential equations
Karl Friedrich Gauss	1799	Fundamental theorem of algebra
Jean Baptiste Joseph Fourier	1807	Fourier series
Augustin-Louis Cauchy	1820	First analysis textbook
Joseph Liouville	1833	A procedure to determine existence of classes of indefinite integrals
Bernhard Bolzano	1834	Developed a nowhere differentiable, everywhere continuous function
Karl Weierstrass	1840	Definition of real numbers, ϵ and δ, "arithmetization of analysis"
G. F. Bernhard Riemann	1854	Riemann sums allow integrals of noncontinuous functions
H. Eduard Heine	1872	The "ϵ-δ" definition of limit in his text *Elements*
Thomas Stieltjes	1894	The Riemann-Stieltjes integral
Henri Lebesgue	1901	Measure and the Lebesgue integral
Robert Risch	1968	The Risch algorithm (a decision procedure to determine an antiderivative's existence)

Figure B.3   Augustin-Louis Cauchy (1789–
1857). Photo by Charles Reutlinger.

Figure B.4   Georg F. B. Riemann (1826–
1866).

Timeline Sources

- Ball, W. W. R. *A Short Account of the History of Mathematics.* (1912)

- Boyer, C. B. *The History of the Calculus and Its Conceptual Development.* (1959)

- Burton, D. M. *The History of Mathematics: An Introduction.* (2007)

- Dunham, W. *Journey Through Genius.* (1990)

- Katz, V. J. and Michelowicz, K. D. *Historical Modules for the Teaching and Learning of Secondary Mathematics (CD).* (2002)

- O'Connor, J. J. and Robertson, E. F. "The MacTutor History of Mathematics Archive." (2008)

APPENDIX C

PROJECTS IN REAL ANALYSIS

Student projects in real analysis can range from simple, expository papers with class presentations to original research on theoretical or pedagogical topics. We present a collection of sample projects for students that begins with historically based writing activities, then moves up through different foci, culminating with Vitali's construction of a nonmeasurable set. Each project can be done in small groups or alone and can be the subject of a classroom student presentation. We close this appendix with links to other sources of student projects for real analysis and to sources of projects for calculus students.

C.1 HISTORICAL WRITING PROJECTS

Fermat, Newton, and Extrema

Describe the methods presented by Fermat, Newton, and a modern calculus text for finding extreme values of a polynomial. Choose a specific third-degree polynomial to illustrate the respective techniques. Create a lesson plan on this topic with supporting material and worksheets for a calculus class.

Introduction to Real Analysis. By William C. Bauldry
Copyright © 2009 John Wiley & Sons, Inc.

Newton versus Leibniz

Describe the controversies over priority that surrounded Newton and Leibniz. Create a lesson plan on this topic with supporting material and worksheets for a calculus class.

Fermat versus Descartes

Describe the controversies over priority that surrounded Fermat and Descartes. Create a lesson plan on this topic with supporting material and worksheets for a calculus class.

Notable Real Analysts

Choose one of the mathematicians whose work we have studied, for example, choose Bernoulli(s), Cauchy, Dirichlet, Euler, Fourier, Gauss, Lebesgue, Riemann, Stieltjes, or Weierstrass, among others. Report on their life, their mathematical work, other interests, the mathematicians they worked with, and so forth. Create a lesson plan on your chosen analyst with supporting material and worksheets for a calculus class.

Leibniz's Harmonic Triangle

Describe *Leibniz's harmonic triangle*. [See, for example, Polya (1962, p. 88).] Compare Leibniz's triangle to *Pascal's triangle*. Create a lesson plan on this topic with supporting material and worksheets for a calculus class.

Birthday Mathematician

Create a worksheet for students outlining the *birthday mathematician project:* Choose a mathematician born on your birthday. Write a brief biography discussing the significance of their work and their historical context. See, e.g., Bauldry & Schellenberg (2006).

C.2   INDUCTION PROOFS: SUMMATIONS, INEQUALITIES, AND DIVISIBILITY

The principle of mathematical induction is one of the strongest and most useful tools in real analysis. And in algebra, and topology, and . . . . We'll start with a summation formula.

Theorem C.1 *For every $n \in \mathbb{N}$,*

$$\sum_{j=1}^{n} 2j = n^2 + n$$

Proof: *Basis.* Let $n = 1$. Then $\sum_{j=1}^{1} 2j = 1^2 + 1$ is true.

Induction. Suppose the formula holds for k. Consider $\sum_{j=1}^{k+1} 2j$. Then

$$\sum_{j=1}^{k+1} 2j = 2(k+1) + \sum_{j=1}^{k} 2j$$
$$= 2(k+1) + (k^2 + k) = (k+1)^2 + (k+1)$$

and the formula holds for $k + 1$.

Conclusion. Therefore, by the principle of mathematical induction, the formula holds for all $n \in \mathbb{N}$. ∎

1. Find and prove a formula for

$$\sum_{j=1}^{n} 2j - 1$$

2. Find and prove a formula for

$$\sum_{j=1}^{n} j^2$$

3. Prove the formula

$$2^0 + 2^1 + 2^2 + 2^3 + 2^4 + \cdots + 2^n = 2^{n+1} - 1$$

Now, let's study an inequality. Observe the beginning of the sequence of values for 2^n and $n!$ in Table C.1.

Table C.1 Values of n, 2^n, and $n!$.

n	2^n	$n!$
1	2	1
2	4	2
3	8	6
4	16	24
5	32	120

We see that $n!$ passes 2^n when $n = 4$ and that $n!$ grows more quickly.

Theorem C.2 *For every natural number n greater than or equal to 4,*

$$2^n < n!$$

Proof: *Basis.* For $n = 4$, we have $16 < 24$.

Induction. Assume that $2^k < k!$ for some integer $k \geq 4$. Then

$$2^{k+1} = 2 \cdot 2^k < 2 \cdot k! < (k+1)!$$

when $k \geq 2$. So the formula holds for $k + 1$.

Conclusion. Therefore, by the principle of mathematical induction, the inequality holds for all $n \in \mathbb{N}$. ∎

4. Prove the inequality $3^n < n!$ for appropriate choice of n.

5. Prove the inequality $4n < 2^n$ for appropriate choice of n.

The last type of induction formula that we will study is for divisibility relationships.

Theorem C.3 *For every $n \in \mathbb{N}$, we have $n^2 - n$ is divisible by 2.*

Proof: *Basis.* For $n = 1$, we have $1^2 - 1 = 0$, which is divisible by 2.

Induction. Assume that $k^2 - k$ is divisible by 2 for some integer k. Then consider $(k + 1)^2 - (k + 1)$.

$$(k + 1)^2 - (k + 1) = (k^2 + 2k + 1) - k - 1$$
$$= (k^2 - k) + (2k)$$

Now, $k^2 - k$ is divisible by 2 by hypothesis and $2k$ is divisible by 2. Hence the sum is divisible by 2; that is, $(k + 1)^2 - (k + 1)$ is divisible by 2.

Conclusion. Therefore, by the principle of mathematical induction, $n^2 - n$ is divisible by 2 for all $n \in \mathbb{N}$. ∎

6. Prove that, for all $n \geq 1$, $8^n - 3^n$ is divisible by 5.

7. Prove that, for all $n \geq 1$, $4^n + 14$ is divisible by 6.

It's instructive to look at an incorrect usage of induction. Consider a very simple example.

Theorem C.4 (False Proposition) *For every $n \in \mathbb{N}$,*

$$\sum_{j=1}^{n} j = \frac{n^2 + n + 1}{2}$$

Proof: *Induction.* Assume for some k that

$$\sum_{j=1}^{k} j = \frac{k^2 + k + 1}{2}$$

Then

$$\sum_{j=1}^{k+1} j = (k + 1) + \sum_{j=1}^{k} j$$

$$= (k + 1) + \frac{k^2 + k + 1}{2} = \frac{k^2 + 3k + 3}{2}$$

$$= \frac{(k + 1)^2 + (k + 1) + 1}{2}$$

Hence the formula holds for $k + 1$.

Conclusion. Therefore, by the principle of mathematical induction, the formula holds for all $n \in \mathbb{N}$. ∎

8. Explain the flaw in the "proof" above.

9. Create a similar example.

We end with Polya's famous example of a "proof" by induction that all horses are the same color.

Theorem C.5 (Polya's Improper Induction) *All horses are the same color.*

Proof: **Basis.** If we have one horse, then it must be of the same color as itself.

Induction. Suppose the result holds for all collections of k horses. Take a collection of $k + 1$ horses. The first k horses have the same color by our induction hypothesis. Ignoring the first horse, the last k horses again have the same color. Since these two collections overlap, then all $k + 1$ horses have the same color.

Conclusion. Therefore, by the principle of mathematical induction, all horses have the same color. ∎

10. Find the flaw Polya designed into the proof.

C.3 SERIES REARRANGEMENTS

A series $\sum a_n$ converges absolutely if and only if the series of absolute values $\sum |a_n|$ converges. If a series $\sum a_n$ converges but $\sum |a_n|$ diverges, then we say $\sum a_n$ *converges conditionally.* A series that converges absolutely may be rearranged, but it will still converge to the same limits. A series that converges conditionally can be rearranged to converge to any number at all. We will rearrange the alternating harmonic series $\sum (-1)^{n+1}/n$ to converge to your favorite number.

Generate your new "favorite number" F by:

Maple randomize(*<birth date>*): F := rand(5 .. 15)()

Mathematica SeedRandom[*<birth date>*]; F := RandomInteger[{5, 15}]

TI-Nspire RandSeed *<birth date>*; F = randInt(5,15)

1. Add positive terms $1/(2k - 1)$ of the alternating harmonic series until the sum is just greater than your favorite number F. Find n_1, the index such that $\sum_{k=1}^{n_1} 1/(2k - 1) > F$. For example, if $F = 16$, then $n_1 = 81897532160124$ terms—a huge number! The harmonic series grows *very* slowly!

2. Add negative terms $-1/(2k)$ to your sum until the total is just under F. Find n_2 such that $\sum_{k=1}^{n_1} 1/(2k - 1) - \sum_{k=1}^{n_2} 1/(2k) < F$. If $F = 16$, then $n_2 = 1$.

3. Add positive terms again until the sum is just greater than F. Find n_3 so that

$$F < \sum_{k=1}^{n_1} \frac{1}{2k-1} - \sum_{k=1}^{n_2} \frac{1}{2k} + \sum_{k=n_1+1}^{n_3} \frac{1}{2k-1}$$

For $F = 16$, we have $n_3 = 222620573466507$.

4. Show why continuing the process will form a series that converges to your favorite number. (Hint: *What is the error in the next term?*)

- Carry out the same process for the series $\sum_{n=2}^{\infty} (-1)^n / \ln(n)$.

- How do the values of n_1, n_2, etc., compare to the earlier calculation?

- Create graphs illustrating the procedure.

C.4   NEWTON AND THE BINOMIAL THEOREM

One of Newton's main tools in integration and in developing his calculus was the binomial series for fractional and negative powers. He started by following Wallis's interpolation procedure but then developed his method for generating the series directly. We'll follow his line of discovery. Newton invented his series before there was a well-defined concept of convergence. We will use computations as Newton did, not proofs, as we proceed.[1]

On June 13, 1676, Newton wrote a letter to England's Royal Society answering Leibniz's request for derivations. Newton's binomial series formula appears in his letter (Newman, 1956, p. 521) as

$$\overline{P + PQ}\Big|\frac{m}{n} = P\frac{m}{n} + \frac{m}{n}AQ + \frac{m-n}{2n}BQ + \frac{m-2n}{3n}CQ + \frac{m-3n}{4n}DQ + \&c.$$

Modern notation for $\overline{P + PQ}\Big|\dfrac{m}{n}$ is $(P + PQ)^{m/n}$. Today, we write this formula as the following proposition.

---

[1] This project is based on a problem set from Polya's *Mathematical Discovery* (1962, p. 91).

Theorem C.6 (Newton's Binomial Theorem) *Let x and $\alpha \in \mathbb{R}$. Then*

$$(1+x)^\alpha = 1 + \frac{\alpha}{1}x + \frac{\alpha(\alpha-1)}{1\cdot 2}x^2 + \frac{\alpha(\alpha-1)(\alpha-2)}{1\cdot 2 \cdot 3}x^3 + \cdots$$

1. Show the series is finite if $\alpha \in \mathbb{N}$.

In a subsequent letter written on October 23, 1767, Newton describes how he began with square roots (Newman, 1956, p. 522). Initially, Newton used Wallis's interpolation method but discovered a pattern in the coefficients and developed his analysis from that.
Suppose that

$$(1+x)^{1/2} = a_0 + a_1 x + a_2 x^2 + a_3 x^3 + \cdots .$$

Then multiplying the series by itself must give $1 + x$ that is,

$$1 + x = (a_0 + a_1 x + a_2 x^2 + a_3 x^3 + \ldots) \times (a_0 + a_1 x + a_2 x^2 + a_3 x^3 + \cdots)$$
$$= a_0^2 + 2a_0 a_1 x + \left(2a_0 a_2 + a_1^2\right)x^2 + 2\left(a_1 a_2 + a_0 a_3\right)x^3 + \cdots$$

So $a_0 = 1$. *Why can't $a_0 = -1$?* The coefficient of x is $2a_0 a_1$, which also equals 1. Hence $a_1 = 1/2$. The remaining coefficients must be zero.

2. Calculate a_2, a_3, and a_4.

3. Is there a pattern to the coefficients?

Let's try the same technique, applying it to $(1+x)^{1/3}$. Suppose that

$$(1+x)^{1/2} = a_0 + a_1 x + a_2 x^2 + a_3 x^3 + \cdots$$

This time we cube the series. As before, a_0 must be 1.

4. Calculate a_1, a_2, a_3, and a_4.

5. Is there a pattern to the coefficients?

6. Can a computer algebra system be used to compute the coefficients?

Newton did not consider convergence. Let a_n be the coefficient of x^n in Theorem C.6.

7. Apply the ratio test (see page 100) to the series.

8. What is the radius of convergence?

Newton's line of reasoning only applied to rational powers $(1+x)^{m/n}$. What would we need to do to extend the result to real powers?

C.5 SYMMETRIC SUMS OF LOGARITHMS

The "Problems and Solutions" sections of journals like the MAA's *Monthly* and the *AMATYC Review* offer more than just fun through challenging questions. The problems also generate good material for the classroom. In the same vein, the fall 2007 issue of the *AMATYC Review* poses an interesting inequality for a symmetric sum of logarithms in four variables. Paraphrased, problem AY–1 (Hjouj & King, 2007, p. 81) asks us to show that, for x, y, z, and $w > 1$,

$$\log_x(xyzw) + \log_y(xyzw) + \log_z(xyzw) + \log_w(xyzw) \geq 16$$

This inequality has an interesting generalization concerning the growth of sums of symmetric logarithms in n variables that only requires a simple minimization and a little term counting to verify.

The Sum of the Logarithms

Begin by minimizing the function

$$f(x, y) = \frac{\ln(x)}{\ln(y)} + \frac{\ln(y)}{\ln(x)}$$

over the region $\{x > 1\} \times \{y > 1\}$.

1. Show this function fails the second derivative test for extrema. [Refer to, for example, Giordano et al. (2005, p. 1030).]

Lemma C.7 *Let* $\mathcal{D} = \{(x, y) \mid x, y > 1\}$. *Define* $f : \mathcal{D} \to \mathbb{R}$ *by*

$$f(x, y) = \frac{\ln(x)}{\ln(y)} + \frac{\ln(y)}{\ln(x)}$$

Then $\min_{(x,y) \in \mathcal{D}} f(x, y) = 2$ *is attained on the line* $\langle t, t \rangle$ *for* $t > 1$.

Proof: Clearly $f(t, t) = 2$ for $t > 1$. Since f fails the second derivative test on the critical line, consider Δf where $\Delta f = f(t + h, t + k) - f(t, t)$.

2. Use a computer algebra system to calculate and simplify Δf.

3. Show that Δf must be nonnegative for all $t + h, t + k > 1$, and so f is minimal on the line $\langle t, t \rangle$ for $t > 1$.
∎

The main inequality is an application of the lemma along with a simple term count. The symmetric sum of the logarithms is greater than the square of the number of indeterminates.

Theorem C.8 *Let $\{x_i\}$ be a collection of n real numbers each greater than 1 and let $P = \prod_{k=1}^{n} x_k$. Then*

$$\sum_{i=1}^{n} \log_{x_i}(P) \geq n^2$$

Proof: Set $P = \prod_{k=1}^{n} x_k$.

4. Show that

$$\log_{x_i}(P) = 1 + \sum_{\substack{k=1 \\ k \neq i}}^{n} \frac{\ln(x_k)}{\ln(x_i)}$$

5. Show the sum can be written as

$$\sum_{i=1}^{n} \log_{x_i}(P) = n + \sum_{i=1}^{n} \sum_{\substack{k=1 \\ k \neq i}}^{n} \frac{\ln(x_k)}{\ln(x_i)}$$

6. Rewrite the sum as

$$\sum_{i=1}^{n} \log_{x_i}(P) = n + \sum_{1 \leq i < k \leq n} \left[\frac{\ln(x_k)}{\ln(x_i)} + \frac{\ln(x_i)}{\ln(x_k)} \right]$$

7. Apply Lemma C.7 to each paired log term on the right side to yield

$$\sum_{i=1}^{n} \log_{x_i}(P) \geq n + \sum_{1 \leq i < k \leq n} 2$$

8. Finish the derivation by counting terms and showing the right side is equal to

$$n + 2 \cdot \binom{n}{2} = n^2$$

∎

Show how the original problem follows immediately from the theorem!

C.6 LOGICAL EQUIVALENCE: COMPLETENESS OF THE REAL NUMBERS

Carefully prove and compare the sequence of results following.

Recall that an *open cover* of a set A is a collection of open sets $\mathcal{O} = \{O_\alpha\}$ such that

$$A \subseteq \bigcup_{O_\alpha \in \mathcal{O}} O_\alpha$$

Definition C.1 (Compact) *A subset K of $\mathbb{R}$ is* compact *if and only if every open cover of K has a finite subcover.*

Give three examples:

1. a compact subset of $\mathbb{R}$

2. a bounded and noncompact subset of $\mathbb{R}$

3. a closed and noncompact subset of $\mathbb{R}$

Now prove the cycle of propositions below.

Theorem C.9 (Completeness of the Real Numbers) *Every nonempty subset of real numbers that has an upper bound has a supremum, or least upper bound.*

Proof: Contradiction by construction.   ∎

Theorem C.10 (Heine-Borel Theorem) *A subset K of $\mathbb{R}$ is* compact *if and only if K is both closed and bounded.*

Proof: Prove each of the following claims.

1. If K is compact, then K is bounded. [Hint: $\mathcal{O} = \{(-n, n) \mid n \in \mathbb{N}\}$ is an open cover.]

2. If K is compact, then K^c is open. [Hint: Let $z \in K^c$. Then $O = \{N_{\delta_x}(x) \mid x \in K\}$ where $\delta_x = |x - z|/2$ is an open cover. Let $\delta = \min\{\delta_x\}$ for δ_x from a finite subcover. Then $N_{\delta_x}(z) \subset K^c$.]

3. Let K be closed and bounded and $\mathcal{O}$ be an open cover. [Hint: Create a contradiction. K is contained in $K_1 = (-n, n)$ for some $n \in \mathbb{N}$. Divide K_1 in half; either the "left" or the "right" half does not have a finite subcover. Repeat the process to get two sequences a_i, the left endpoints, and b_i, the right endpoints. Both converge to the same point k using completeness. Show $k \in K$ since K is closed and $k \notin K$ by construction.]

                                                                              ∎

Theorem C.11 (Bolzano-Weierstrass Theorem) *Every bounded infinite set of real numbers has an accumulation point.*

Proof: Suppose an infinite set is bounded but has no accumulation points. Apply the Heine-Borel theorem.

  If there are no accumulation points, then every point x in the set has a neighborhood $N_{r_x}(x)$ such that the neighborhood's intersection with the set is equal to only the point x. What is the relation between the collection $\mathcal{O} = \{N_{r_x}(x) \mid x \text{ is in the set}\}$ and the set itself? Generate a contradiction to the assumption that the set has no accumulation points.   ∎

Theorem C.12 (Bounded Monotone Sequence Theorem) *Every bounded monotone sequence of real numbers has a limit.*

Proof: Assume $X = \{x_n\}$ is a bounded, increasing sequence. If X is finite, then the result is easy. Suppose X is infinite. Apply the Bolzano-Weierstrass theorem to find an accumulation point. Show this point is the limit of the sequence.   ∎

Complete the circle by now proving

Theorem C.13 (Completeness of the Real Numbers) $\mathbb{R}$ *is complete.*

Proof: Let A be a bounded, nonempty set of real numbers. Let b be any upper bound of A. Choose any point $x_1 \in A$. If x_1 is the supremum, we're done. If not, set $x_2 = (x_1 + b)/2$. If x_2 is the supremum, we're done. Continue this process to create a bounded, infinite sequence. Show the bounded monotone sequence theorem applies.   ∎

We have proved: *Completeness of* $\mathbb{R}$ $\implies$ *Heine-Borel Theorem* $\implies$ *Bolzano-Weierstrass Theorem* $\implies$ *Bounded Monotone Sequence Theorem* $\implies$ *Completeness of* $\mathbb{R}$. Since going around the circle of implications shows any result yields any other, the implications are all really 'if and only if.' What does this imply about the four results?

Finish a presentation with an example showing that $\mathbb{Q}$ is *not* complete.

C.7   VITALI'S NONMEASURABLE SET

Giuseppe Vitali was the first to construct a set that was non-Lebesgue measurable. His construction is based on countable additivity, a critical property of measures. Watch for the axiom of choice to appear!

Let $X = [0, 1)$ and define the operation $\oplus : X \to X$ by addition modulo 1. Then $x \oplus y = x + y - \lfloor x + y \rfloor$. (Note: X is then equivalent to the unit circle via $t \mapsto e^{it}$.) Define addition of a set with a number by $A \oplus x = \{a \oplus x \mid a \in A\}$.

Now define the relation $\sim$ on X by

$$x \sim y \text{ if and only if there is a rational } r \text{ such that } |x - y| = r$$

1. Show that for each rational $r \in \mathbb{Q} \cap X$ we have $r \sim 0$, and so all rationals are equivalent under $\sim$.

2. Prove that $\sim$ is an equivalence relation on $X = [0, 1)$. Let $[x]$ be *the equivalence class of* x, that is, $[x] = \{y \in X \mid y \sim x\}$.

3. Find $[0]$.

Consider $x_1 = \pi/10$ and $x_2 = \pi/30$. Since $x_1 - x_2 = \pi/15 \notin \mathbb{Q}$, then $[x_1] \neq [x_2]$. Now take $x_3 = (\pi + 5)/10$. Since $x_1 - x_3 = 1/2 \in \mathbb{Q}$, we have $x_1 \sim x_3$, so $[x_1] = [x_3]$.

Since $\sim$ is an equivalence relation, it partitions X. Choose one representative h from each equivalence class in the partition of X. (*Axiom of Choice!*) Gather the representative h to form the set H. Consider the collection of these sets $\mathcal{H} = \{H \oplus r\}$ where r ranges over the rationals in X.

4. Determine whether H is countable or uncountable.

5. Verify that $\mathcal{H}$ is a pairwise-disjoint family; i.e., $(H \oplus r_1) \cap (H \oplus r_2) = \emptyset$ for $r_1 \neq r_2$.

6. Prove that

$$X = \bigcup_{r \in \mathbb{Q} \cap X} (H + r)$$

Since Lebesgue measure is translation invariant, $\mu(H \oplus r) = \mu(H)$ for all $r \in \mathbb{Q} \cap X$. Assume that H is Lebesgue measurable, and $\mu(H) = \lambda$. Then, since $\mathcal{H}$ is a countable family of disjoint sets,

$$1 = \mu(X) = \mu \left(\bigcup_{r \in \mathbb{Q} \cap X} (H + r) \right) = \sum_{r \in \mathbb{Q} \cap X} \lambda$$

We have a contradiction: If $\lambda = 0$, then $1 = 0$. Otherwise, if $\lambda > 0$, then $1 = \infty$. Thus H cannot be Lebesgue measurable.

This construction is due to Vitali (1905). *Even though Vitali's paper is written in Italian, enough of the detail is "in mathematics" that we can understand the paper without being able to read Italian. Try it!*

C.8   SOURCES FOR REAL ANALYSIS PROJECTS

A collection of sources for projects for students of real analysis follows.

- Brabenec's *Resources for the Study of Real Analysis* (2006)

- Snow and Weller's *Exploratory Examples for Real Analysis* (2007)

- Kosmala's *A Friendly Introduction to Analysis* (2004): each chapter ends with a collection of student projects.

- Shakarchi and Lang's *Problems and Solutions for Undergraduate Analysis* (1997)

- Aliprantis and Burkinshaw's *Problems in Real Analysis: A Workbook with Solutions* (1999) (advanced real analysis)

C.9   SOURCES FOR PROJECTS FOR CALCULUS STUDENTS

There are many books and websites containing projects and laboratory exercises for calculus classes. A small selection follows.

- Bauldry and Fiedler's *Calculus Projects with Maple* (1996)

- Crannell, LaRose, and Ratliff's *Writing Projects for Mathematics Courses: Crushed Clowns, Cars & Coffee to Go* (2004)

- Gaughan et al's *Student Research Projects in Calculus* (1991)

- Packel and Wagon's *Animating Calculus: Mathematica Notebooks for the Laboratory* (1997)

- Solow and Fink's *Learning by Discovery: A Lab Manual for Calculus* (1993)

- Stroyan's *Projects for Calculus: The Language of Change* (1998)

- Wood's *Calculus Mysteries and Thrillers* (1999)

Many calculus texts offer projects following the chapters.

- *Calculus: Single and Multivariable*, by Hughes-Hallett et al. (2009)

- "Calculus: Modeling and Application," online text by Moore & Smith (2004)

- *Calculus from Graphical, Numerical, and Symbolic Points of View*, by Ostebee & Zorn (2002)

- *Calculus*, by Stewart (2009)

The last word:

Ancora imparo.   *—Michelangelo*

BIBLIOGRAPHY

Abramowitz, M. & Stegun, I. *Handbook of Mathematical Functions with Formulas, Graphs, and Mathematical Tables*. Dover, New York. 1965. Available online at http://www.math.sfu.ca/~cbm/aands/frameindex.htm.

Aigner, M. & Ziegler, G. M. *Proofs from THE BOOK*. Springer, New York, 2nd edition. 2001.

Aliprantis, C. D. & Burkinshaw, O. *Problems in Real Analysis: A Workbook with Solutions*. Academic Press, San Diego, CA. 1999.

Anton, H., Bivens, I. C., & Davis, S. *Calculus Early Transcendentals Single Variable*. John Wiley & Sons, New York, 9th edition. 2009.

Apostol, T. M. *Calculus*, volume II. John Wiley & Sons, New York, 2nd edition. 1969.

Ball, W. W. R. *A Short Account of the History of Mathematics*. Macmillan, London, 5th edition. 1912. Facsimile edition, Sterling, New York. 2001.

Bardi, J. S. *The Calculus Wars*. Thunder's Mouth Press, New York. 2006.

Barnard, T. "Why Are Proofs Difficult?" *The Mathematical Gazette*, **84**(501), 415–422. 2000.

Bartle, R. G. *The Elements of Integration and Lebesgue Measure*. Wiley Classics Library. John Wiley & Sons, New York. 1995.

Bauldry, W. C. "Fitting Logistics to the U.S. Population." *MapleTech*, **4**(3), 73–77. 1997.

Bauldry, W. C. & Fiedler, J. R. *Calculus Projects with Maple*. Brooks/Cole, Pacific Grove, CA, 2nd edition. 1996.

Bauldry, W. C. & Schellenberg, D. "A Note on the 'Mathematician's Birthday Calendar'." *NCCTM Centroid*, **32**(2), 6. 2006.

Bear, H. S. *A Primer of Lebesgue Integration.* Academic Press, San Diego, CA. 1995.

Bhatia, R. *Fourier Series.* Classroom Resource Materials. Math. Assoc. of America, Washington, DC. 2005.

Bichteler, K. *Integration – A Functional Approach.* Birkhäuser, New York. 1998.

Boas, R. P. *A Primer of Real Functions.* The Carus Mathematical Monographs. Math. Assoc. of America, Washington, DC. 1981.

Bonar, D. D. & Khoury, M. J. *Real Infinite Series.* Classroom Resource Materials. Math. Assoc. of America, Washington, DC. 2006.

Bonnet, P. O. "Rémarques sur quelques intégrales définies." *J. Math. Pures Appl.*, **14**, 249–256. 1849.

Borzellino, J. E. "Whose Limit Is It Anyway?" *PRIMUS*, **XI**(3), 265–274. 2001.

Boyer, C. B. "Fermat's Integration of X^n." *National Mathematics Magazine*, **20**(1), 29–32. 1945.

Boyer, C. B. *The History of the Calculus and Its Conceptual Development.* Dover, New York. 1959.

Brabenec, R. L. *Resources for the Study of Real Analysis.* Classroom Resource Materials. Math. Assoc. of America, Washington, DC. 2006.

Bressoud, D. M. *A Radical Approach to Real Analysis.* Classroom Resource Materials. Math. Assoc. of America, Washington, DC. 2005.

Bressoud, D. M. *A Radical Approach to Lebesgue's Theory of Integration.* MAA Textbooks. Cambridge University Press, Washington, DC. 2008.

Brown, J. "Mathematics, Physics and *A Hard Day's Night*." *CMS Notes*, **36**(6), 4–8. 2004.

Burk, F. E. *A Garden of Integrals.* Number 31 in Dolciani Mathematical Expositions. Math. Assoc. of America, Washington, DC. 2007.

Burton, D. M. *The History of Mathematics: An Introduction.* McGraw-Hill, New York, 6th edition. 2007.

Calkin, N. & Wilf, H. S. "Recounting the Rationals." *The American Mathematical Monthly*, **107**(4), 360–363. 2000.

Cauchy, A.-L. *Cours d'Analyse de l'École Royale Polytechnique.* Chez Debure frères, Paris. 1821. Available online at http://gallica.bnf.fr/ark:/12148/bpt6k29058v.

Chae, S. B. *Lebesgue Integration.* Springer-Verlag, New York, 2nd edition. 1995.

Cohn, D. L. *Measure Theory.* Birkhäuser, New York. 1980.

Coolidge, J. L. "The Story of Tangents." *The American Mathematical Monthly*, **58**(7), 449–462. 1951.

Crannell, A., LaRose, G., & Ratliff, T. *Writing Projects for Mathematics Courses: Crushed Clowns, Cars & Coffee to Go.* Classroom Resource Materials. Math. Assoc. of America, Washington, DC. 2004.

Davidson, K. R. & Donsig, A. P. *Real Analysis with Real Applications.* Prentice Hall, Upper Saddle River, NJ. 2002.

Davis, P. J. "Leonhard Euler's Integral: A Historical Profile of the Gamma Function." *The American Mathematical Monthly*, **66**(10), 849–869. 1959.

Davis, W., Portia, H., & Uhl, J. *Calculus & Mathematica.* Addison-Wesley, Reading, MA. 1994.

Dehn, M. & Hellinger, E. D. "Certain Mathematical Achievements of James Gregory." *The American Mathematical Monthly*, **50**(3), 149–163. 1943.

Diaconis, P. & Freedman, D. "An Elementary Proof of Stirling's Formula." *The American Mathematical Monthly*, **93**(2), 123–125. 1986.

Duc, N. M. "Farmers' Satisfaction with Aquaculture—A Logistic Model in Vietnam." *Ecological Economics*, **68**(1-2), 525–531. 2008.

Dunham, W. *Journey Through Genius*. John Wiley & Sons, New York. 1990.

Dunham, W. *Euler: The Master of Us All*. Number 22 in Dolciani Mathematical Expositions. Math. Assoc. of America, Washington, DC. 1999.

Dunham, W. *The Calculus Gallery: Masterpieces from Newton to Lebesgue*. Princeton University Press, Princeton, NJ. 2008.

Dunnington, G. W. *Gauss: Titan of Science*. Math. Assoc. of America, Washington, DC. 2004.

Ellis, W., Bauldry, W., Fiedler, J., Giordano, F., Judson, P., Lodi, E., Vitray, R., & West, R. *Calculus: Mathematics and Modeling*. Addison-Wesley, Reading, MA, revised preliminary edition. 1999.

Farand, S. M. & Poxon, N. J. *Calculus*. Harcourt Brace College Outline Series. Harcourt, San Diego, CA. 1984.

Finney, R., Demana, F., Waits, B., & Kennedy, D. *Calculus*. Addison-Wesley, Reading, MA. 1999.

Fourier, J. *The Analytical Theory of Heat*. Dover, New York. 1822, 2003. Unabridged republication of the 1878 translation.

Gaughan, E. D., Pengelley, D. J., Knoebel, A., & Kurtz, D. *Student Research Projects in Calculus*. Math. Assoc. of America, Washington, DC. 1991.

Giordano, F. R., Hass, J., & Weir, M. D. *Thomas' Calculus*. Addison-Wesley, Reading, MA. 2005.

Grabiner, J. V. "The Changing Concept of Change," chapter in Jahnke, H. N. (ed.) *Sherlock Holmes in Babylon and Other Tales of Mathematical History*, pp. 218–227. Math. Assoc. of America, Washington, DC. 2004.

Gradshteyn, I. S. & Ryzhik, I. M. *Table of Integrals, Series, and Products*. Academic Press, San Diego, CA, 7th edition. 2007.

Granville, W. A., Smith, P. F., & Longley, W. R. *Elements of the Differential and Integral Calculus*. Ginn and Co., Boston. 1911.

Hairer, E. & Wanner, G. *Analysis by Its History*. Springer-Verlag, New York. 1996.

Hawkins, T. *Lebesgue's Theory of Integration: Its Origins and Development*. AMS Chelsea Pub./Amer Math. Soc., New York, 2nd edition. 2002.

Heide, T. "History of Mathematics and the Teacher," chapter in Calinger, R. (ed.) *Vita Mathematica: Historical Research and Integration with Teaching*, pp. 241–243. Number 40 in MAA Notes. Math. Assoc. of America, Washington, DC. 1996.

Herman, R. A., Scherer, P. N., & Shan, G. "Evaluation of Logistic and Polynomial Models for Fitting Sandwich-ELISA Calibration Curves." *Journal of Immunological Methods*, **339**(2), 245–258. 2008.

Hjouj, F. & King, R. "Problem AY-1." *AMATYC Review*, **29**(1), 81. 2007.

Hochkirchen, T. "Theory of Measure and Integration from Riemann to Lebesgue," chapter in Jahnke, H. N. (ed.) *Sherlock Holmes in Babylon and Other Tales of Mathematical History*, pp. 261–290. Math. Assoc. of America, Washington, DC. 2004.

Hughes-Hallett, D., Gleason, A. M., McCallum, W. G., Flath, D. E., Lock, P. F., Tucker, T. W., Lomen, D. O., Lovelock, D., Mumford, D., Osgood, B. G., Quinney, D., Rhea, K., & Tecosky-Feldman, J. *Calculus: Single and Multivariable.* John Wiley & Sons, New York, 5th edition. 2009.

Jackson, D. *Fourier Series and Orthogonal Polynomials.* Carus Monographs. Math. Assoc. of America, Washington, DC. 1941. Dover edition, Dover, 2004.

Kahane, J.-P. "A Century of Interplay Between Taylor Series, Fourier Series and Brownian Motion." *Bulletin of the London Mathematical Society*, **29**(03), 257–279. 2000.

Kasper, T. "Integration in Finite Terms: The Liouville Theory." *Mathematics Magazine*, **53**(4), 195–201. 1980.

Katz, V. J. & Michelowicz, K. D. *Historical Modules for the Teaching and Learning of Secondary Mathematics (CD).* Math. Assoc. of America, Washington, DC. 2002.

Kleiner, I. "Evolution of the Function Concept: A Brief Survey." *The College Mathematics Journal*, **20**(4), 282–300. 1989.

Körner, T. W. *Fourier Analysis.* Cambridge University Press, Cambridge, UK. 1989.

Kosmala, W. A. J. *A Friendly Introduction to Analysis.* Prentice Hall, Upper Saddle River, NJ, 2nd edition. 2004.

Kuhn, S. "The Derivative a la Carathéodory." *The American Mathematical Monthly*, **98**(1), 40–44. 1991.

Lagrange, J. L. *Théorie Des Fonctions Analytiques.* Imprimerie de la République, Paris. 1797. Available online at http://books.google.com/books?id=15IKAAAAYAAJ.

Littlewood, J. E. *Lectures on the Theory of Functions.* Oxford Univ. Press, London. 1944.

Ma, L. *Knowing and Teaching Elementary Mathematics.* Lawrence Erlbaum Associates, Mahwah, NJ. 1999.

Malthus, T. R. *An Essay on the Principle of Population.* John Murray, London, 6th edition. 1826. Available online at http://www.econlib.org/library/Malthus/malPlong.html.

Marchisotto, E. A. & Zakeri, G.-A. "An Invitation to Integration in Finite Terms." *The College Mathematics Journal*, **25**(4), 295–308. 1994.

Michener, E. R. "Understanding Understanding Mathematics." *Cognitive Science*, **2**, 361–383. 1978.

Moore, L. & Smith, D. "Calculus: Modeling and Application." 2004. Available online at http://www.math.duke.edu/education/calculustext/.

Moritz, R. E. *On Mathematics and Mathematicians.* Dover, New York. 1958. Available online at http://ia331305.us.archive.org/3/items/onmathematicsand017018mbp/ onmathematicsand017018mbp.pdf.

Newman, J. R. *The World Of Mathematics*, volume 1. Simon & Schuster, New York. 1956.

O'Connor, J. J. & Robertson, E. F. "The MacTutor History of Mathematics Archive." 2008. Available online at http://www-gap.dcs.st-and.ac.uk/~history/.

Ostebee, A. & Zorn, P. *Calculus from Graphical, Numerical, and Symbolic Points of View.* Brooks/Cole, Pacific Grove, CA, 2nd edition. 2002.

Packel, E. W. & Wagon, S. *Animating Calculus: Mathematica Notebooks for the Laboratory.* TELOS, Santa Clara, CA. 1997.

Pearl, R., Reed, L. J., & Kish, J. F. "The Logistic Curve and the Census Count of 1940." *Science, New Series*, **92**(2395), 486–488. 1940.

Peleg, M., Corradini, M. G., & Normand, M. D. "The Logistic (Verhulst) Model for Sigmoid Microbial Growth Curves Revisited." *Food Research International*, **40**(7), 808–818. 2007.

Polya, G. *Mathematical Discovery*, volume 1. John Wiley & Sons, New York. 1962.

Rainville, E. D. *Special Functions*. Chelsea, New York. 1971.

Richardson, D. "Some Undecidable Problems Involving Elementary Functions of a Real Variable." *The Journal of Symbolic Logic*, **33**(4), 514–520. 1968.

Risch, R. H. "The Problem of Integration in Finite Terms." *Transactions of the American Mathematical Society*, **139**, 167–189. 1969.

Royden, H. *Real Analysis*. Macmillan, New York, 3rd edition. 1988.

Rudin, W. *Principles of Mathematical Analysis*. McGraw-Hill, New York, 3rd edition. 1976.

Shakarchi, R. & Lang, S. *Problems and Solutions for Undergraduate Analysis*. Springer, New York. 1997.

Shulman, B. "Math-Alive! Using Original Sources To Teach Mathematics in Social Context." *PRIMUS*, **8**(1), 1–14. 1998.

Snow, J. E. & Weller, K. E. *Exploratory Examples for Real Analysis*. Classroom Resource Materials. Math. Assoc. of America, Washington, DC. 2007.

Solow, A. E. & Fink, J. B. *Learning by Discovery*. Number 27 in MAA Notes. Math. Assoc. of America, Washington, DC. 1993.

Stewart, J. *Calculus*. Thomson Brooks/Cole, Pacific Grove, CA, 6th edition. 2009.

Stillwell, J. *Mathematics and Its History*. Springer-Verlag, New York. 1989.

Stroyan, K. D. *Projects for Calculus: The Language of Change*. Academic Press, San Diego, CA. 1998.

Struik, D. J. (ed.). *A Source Book in Mathematics, 1200–1800*. Princeton University Press, Princeton, NJ. 1986.

Thomas, G. B. *Calculus and Analytic Geometry*. Addison-Wesley, Reading, MA, 4th edition. 1968.

Underwood, N. "Variation in and Correlation between Intrinsic Rate of Increase and Carrying Capacity." *The American Naturalist*, **169**(1), 136–141. 2007.

Verhulst, P. F. "Notice sur la loi que la population poursuit dans son accroissement." *Correspondance mathématique et physique*, **X**, 113–121. 1838. English translation by Vogels et al, *J. Biol. Phys*. 3, 183–192, 1975.

Verhulst, P. F. "Recherches mathématiques sur la loi d'accroisement de la population." *Mem. Acad. R. Bruxelles*, **18**, 1–58. 1844.

Vitali, G. *Sul problema della misura dei gruppi di punti di una retta*. Tip. Gamberini e Parmeggiani, Bologna, IT. 1905.

Woods, R. G. *Calculus Mysteries and Thrillers*. Classroom Resource Materials. Math. Assoc. of America, Washington, DC. 1999.

Wrede, R. C. & Spiegel, M. R. *Schaum's Outline of Advanced Calculus*. Schaum's Outlines. McGraw-Hill, New York. 2002.

Zwillinger, D. *Handbook of Integration*. Jones and Bartlett, Boston. 1992.

Zwillinger, D. (ed.). *CRC Standard Mathematical Tables and Formulæ*. CRC Press, Boca Raton, FL, 31st edition. 2002.

INDEX